UNION INTERNATIONALE DES SCIENCES PRÉHISTORIQUES ET PROTOHISTORIQUES
INTERNATIONAL UNION OF PREHISTORIC AND PROTOHISTORIC SCIENCES

PROCEEDINGS OF THE XVI WORLD CONGRESS (FLORIANÓPOLIS, 4-10 SEPTEMBER 2011)
ACTES DU XVI CONGRÈS MONDIAL (FLORIANÓPOLIS, 4-10 SEPTEMBRE 2011)

VOL. 9

uispp

Mobility and Transitions in the Holocene

Edited by

Luiz Oosterbeek
Cláudia Fidalgo

BAR International Series 2658
2014

Published in 2016 by
BAR Publishing, Oxford

BAR International Series 2658

Mobility and Transitions in the Holocene

ISBN 978 1 4073 1300 9

Proceedings of the XVI World Congress of the International Union of Prehistoric and Protohistoric Sciences
Actes du XVI Congrès mondial de l'Union Internationale des Sciences Préhistoriques et Protohistoriques
Secretary of the Congress: Rossano Lopes Bastos
President of the Congress National Commission: Erika Robrhan-Gonzalez
Elected President: Jean Bourgeois
Elected Secretary General: Luiz Oosterbeek; Elected Treasurer: François Djindjian
Series Editors: Luiz Oosterbeek, Erika Robrhan-Gonzalez
Volume title: Mobility and Transitions in the Holocene
Volume editors: Luiz Oosterbeek and Cláudia Fidalgo

The signed papers are the sole responsibility of their authors.
Les textes signés sont de la seule responsabilité de leurs auteurs.

Contacts: General Secretariat of the U.I.S.P.P. – International Union of Prehistoric and Protohistoric Sciences,
Instituto Politécnico de Tomar, Av. Dr. Cândido Madureira 13, 2300 TOMAR, Email: uispp@ipt.pt

BAR Publishing is the trading name of British Archaeological Reports (Oxford) Ltd.
British Archaeological Reports was first incorporated in 1974 to publish the BAR
Series, International and British. In 1992 Hadrian Books Ltd became part of the BAR
group. This volume was originally published by Archaeopress in conjunction with
British Archaeological Reports (Oxford) Ltd / Hadrian Books Ltd, the Series principal
publisher, in 2014. This present volume is published by BAR Publishing, 2016.

Printed in England

BAR
PUBLISHING

BAR titles are available from:

BAR Publishing
122 Banbury Rd, Oxford, OX2 7BP, UK
EMAIL   info@barpublishing.com
PHONE   +44 (0)1865 310431
FAX   +44 (0)1865 316916
www.barpublishing.com

Table of Contents

List of Figures and Tables

M. OTTE: **Substrats néolithiques aux arts traditionnels des Balkans**

J. DE OLIVEIRA, C. OLIVEIRA: Other faces of the Megalithic in the north-east Alentejo, Portugal, and the reuse of tombs

L. ROCHA, R. FERNANDES: Some possible assessments of the different burial spaces in the Alentejo and Arrábida in prehistory and protohistory

C. LAZĂR: The eneolithic necropolis from Sultana-Malu Rosu (Romania) – a case study

E. GÓMEZ: A group of offerings of Cotzumalguapa, Guatemala: late classic period

Introduction

The organization of the UISPP XVI world congress in Florianópolis was the occasion to focus a certain number of themes that are preferably dealt with at a transcontinental scale. Several session discussed the issue of transition mechanism (technological, social, economic, and their climatic and environmental contexts).

Marcel Otte opens the volume, focusing on the specific role of straits, a topic that is also at the foundation of Judith Carlin's *et al.* paper.

Contribution by Fabio Parenti *et al.*, Gustavo Wagner and Mercedes Okumura *et al.*, discuss the human adaptations in different contexts in Brazil, during the early and middle Holocene.

First farming societies in Southern America and in Europe are approached in the papers by Marcel Otte and Jorge Oliveira *et al.*, while the transition into more complex societies, bearing mettalurgical knowledge, is the focus of papers by Leonor Rocha *et al.*, Cătălin Lazăr.

Finally, classic contexts on both sides of the Atlantic are revisited by Erika Gómez and by Carolina Dias.

Luiz Oosterbeek

DÉTROITS

Marcel OTTE

Professeur de Préhistoire, Université de Liège, Belgique

Abstract: *Straits are challenges for the human spirit, which has unceasingly confronted and crossed them during all time periods. The real question is why some crossings did not take place en masse all of the time.*

Key-words: *Straights – Mobility*

Résumé: *Les détroits constituent des défis pour l'esprit humain qui n'a jamais cessé de les relever à toutes époques. La véritable question se place dans la raison pour laquelle certains passages en masse ne furent pas exécutés en tout temps.*

Mots-clés: *Détroits – Mobilité*

Imaginez. D'autres hommes habitent sur la terre ferme, et n'en sortent pas! Nous avons toujours vécu sur la mer, ses détroits, ses estuaires grâce à sa prodigalité permanente. Par d'anciens récits, nous savons que d'autres ont craint cette mer hostile, redoutable à leurs yeux, pourtant providentielle, pour ses ressources autant que pour ses contacts. Curieuse appréhension, qui nous fit voir exactement à l'envers toutes ces valeurs aux fondements de nos existences, ont forgé nos mythes. Nos rythmes suivent les cycles des saisons, guidés par la houle des mers, par les mouvements des astres auxquels nous nous sentons perpétuellement liés, de notre naissance à notre destin. Ainsi, notre vie fut conçue, rythmée, agrémentée, éternelle, comme la mer elle-même.

Pour les hommes restés à terre, ces passages constants, cette vie marine, peuvent s'illustrer par d'abondants témoignages que les battements des mers préservèrent sur leurs marges. Par exemple, les obsidiennes, belles, précieuses et fragiles, traversèrent les deux branches de la « mer intérieure » au sein même de notre continent, de la Sicile à la Tunisie actuelles. Les hauts-fonds, alors exondés, établissaient ce passage, en une sorte de vie marine dont les ressources furent à la fois exploitées, échangées et aux sources de nos mythes, de part et d'autre de cet estuaire (fig. 1). Les symboles traversent les détroits en de multiples reprises, comme les hommes qui les portent, sans pouvoir clairement distinguer les uns des autres: les influences mythiques furent portées autant par les peuples que par leurs valeurs.

Le plus évident, pour les peuples de la terre, fut porté par le symbole bifacial, présent partout mais jamais exploité avec de tels succès, dès les périodes les plus anciennes, depuis 650 millénaires de l'Afrique à l'Italie (Venosa: Piperno, 1999).

utant la technique fut complexe et cohérente, autant la mythologie fut élaborée sur des fondements idéologiques totalement autonomes. Ce que nous pouvons en savoir par les traces incarnées dans les images montre en effet une relation toute différente entre l'esprit des hommes et ceux de l'univers. Cette relation fut allusive selon la voie des mythes en préhistoire européenne, mais elle fut narrative dans le monde africain (fig. 2), celui-là même qui va dicter notre « langage des formes » jusqu'à la Renaissance. Toujours, depuis lors, l'agencement des images a suivi celui du récit verbal: jamais avant, plus jamais ensuite. Mais dès que nous touchons à l'histoire des formes, nous sommes à la fois tenus et guidés par les autres aventures suivies par les restes de pensées plastiques auxquels celles-ci furent liées (fig. 2) et comme accrochées à leurs limites continentales.

Voyons quels en furent les grands éléments qui les ont formées au-delà des simplifications, souvent réductrices, qu'ont fait subir les préhistoriens, et eux seuls, à l'histoire des arts plastiques. Cette situation nous permet et nous oblige à considérer l'histoire des formes comme le reflet, parmi tant d'autres, de l'histoire des civilisations. Or, nous voyons se déplacer des mouvements d'idées, comme des mouvements de peuples, au travers de couloirs marins, inconcevables aujourd'hui, au moins par deux catégories de raisons. La première tient aux battements des niveaux marins, et des mouvements des fluides qui furent suivis. La seconde, beaucoup plus puissante et plus mystérieuse, tient aux dogmes, souvent terrestres, qui ont formé notre pensée. Quel navigateur, aussi aventureux que Christophe Colomb aurait imaginé d'aussi denses réseaux de liaisons, réunissant, à travers la haute mer, des points aussi isolés que les îles du Pacifique, à travers tous les temps et aux époques les plus variées?

Le cas des côtes siciliennes fut le plus éclairant de ces évidences « aveuglantes ». Non seulement les destins y sont semblables à ceux de la Tunisie toute proche, par les techniques et les styles employés de part et d'autre du détroit, mais surtout les représentations graphiques s'inspirent du même fond narratif, totalement étranger aux « mythogrammes » européens (Leroi-Gourhan, 1965). Par ailleurs, les hauts-fonds, alors exondés, établissaient comme une plateforme continentale reliant

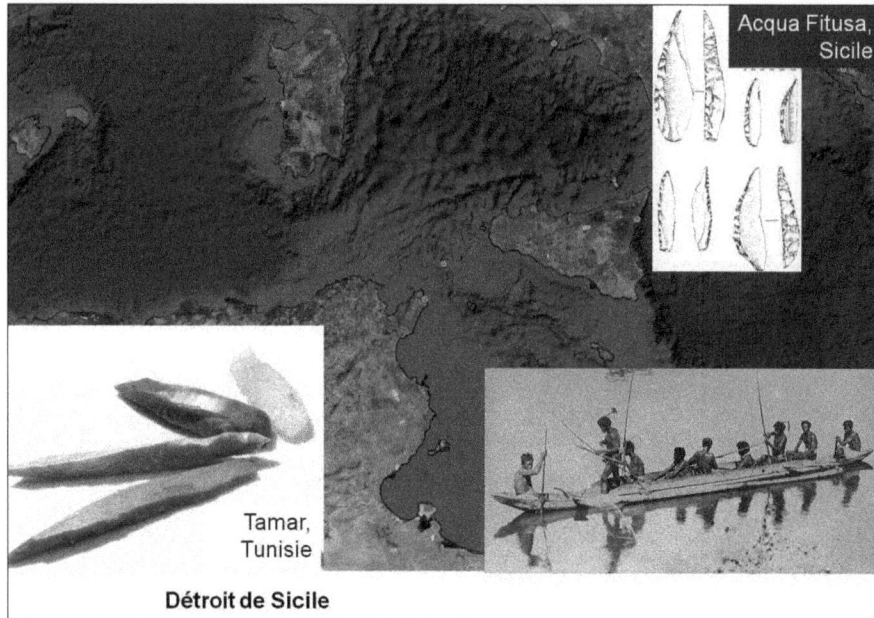

Figure 1. En basses eaux (grisé clair), le détroit de Sicile fut facilement franchi, comme les modes techniques et les matériaux le prouvent. À mesure du recul marin, les territoires parcourus furent marins, mais les contacts persistèrent via la navigation, autant comme moyen de déplacement que comme source alimentaire

Figure 2. Les modes d'expression plastique témoignent, mieux que d'autres, de la communauté d'inspiration mythologique et graphique. Les relations entretenues sous forme de scènes furent directement dérivées des traitements africains des images

les deux ailes du détroit. Ainsi, ni les uns ni les autres ne participaient plus aux mondes originels d'où pourtant ils furent issus, mais ils créèrent une sorte de monde particulier, à substrat maritime, et autant fondé sur chacune des deux rives des continents opposés. Cette illustrations en fut donnée dès l'Acheuléen le plus ancien et à travers toutes les phases intermédiaires de la préhistoire.

De telle sorte que, lors des phases glaciaires, seul un mince bras de mer séparait les deux mondes, encore fut-il réduit à un simple lac allongé dans les périodes les plus exondées. Ainsi, les techniques furent identiques de chaque côté de cette « plaine maritime », les échanges mythiques et artistiques participaient aux mêmes croyances, mais encore notre territoire était marin à l'instar d'innombrables peuples actuels. Une extension de la démographie africaine faisait suite aux nombreuses autres, depuis l'Acheuléen. L'art y était spécialement particulier par ses « mises en scène », ses personnages masqués et, surtout, l'omniprésence humaine, en complète opposition avec tout l'art connu en Eurasie, plus

Figure 3. Les îles, intermédiaires entre les deux continents, surgissent dès l'abaissement du niveau marin. Les distances à traverser furent alors d'autant plus courtes que les plages furent beaucoup plus larges qu'aujourd'hui. Les témoins de passages furent innombrables mais, curieusement, sporadiques. L'Acheuléen constitue l'un des plus spectaculaires exemples de ces contacts

tôt ou plus tard (fig. 2). La liaison intellectuelle entre les personnages constituait la véritable signification de telles scènes, et non plus leurs valeurs de rêves mythiques.

Le même phénomène se présente à Gibraltar, tout proche: deux masses démographiques se font face, chacune liée à un immense arrière-pays, pratiquement sans limite. Cependant, un pont terrestre ne fut jamais installé pendant la période humaine bien que des chapelets d'îles ont émergé à plusieurs reprises à mesure où les niveaux marins descendaient lors des phases glaciaires (fig. 3). Quelques kilomètres seulement les séparaient, soit entre, soit avec les rivages, dont les côtes furent en outre rapprochées par les plages largement exondées. En réalité, le cas de Gibraltar se présente de façon aussi paradoxale qu'inverse par rapport aux modes de pensées dominantes aujourd'hui. Comment, en effet, de si proches côtes n'ont-elles pas produit des contacts beaucoup plus fréquents et plus denses que la littérature archéologique tend à le laisser croire? Spécialement, lorsqu'on observe l'autre rive, si imposante de quelque côté où l'on se place. En effet, si la vue en mer se limite à environ 12 km par sa convexité, dès la plus minime ascension, cette vision lointaine s'accroît extrêmement vite: une simple expérience visuelle le prouve avec abondance. L'esprit humain fut toujours attiré par ces performances (encore actuellement). Or, tout le paléolithique moyen n'en montre pas la trace, sauf au tout début (Acheuléen) et en toute fin (Atérien) (fig. 4). Seule, une barrière de caractère sociologique (la crainte de l'autre) ou mythique (la crainte d'autres territoires) a pu l'expliquer. Pourtant, ces appréhensions se sont brisées à plusieurs reprises et seules celles des archéologues ont persisté. Par exemple, le passage de l'Acheuléen (1 million d'années, au moins, à Casablanca) s'est effectué en masse vers l'Ibérie, mais

très tardivement (500.000 ans à Ambrona). Gibraltar a donc fonctionné tel un filtre plutôt que comme une barrière, d'autant que les industries archaïques sont connues à l'extrême sud de l'Espagne (Orce), datées de plus d'un million d'années, mais sans aucun rapport avec les innovations contemporaines, voire plus anciennes, en Afrique (1,6 millions d'années pour l'Acheuléen au Kenya). Plus que les traversées attestées, leurs absences si longues posent un problème. Elles réapparaissent avec régularité à partir d'environ 20.000 ans, lorsque les pièces bifaciales de l'Atérien récent viennent briser l'évolution des traditions de l'extrême sud-ouest de l'Eurasie où elles s'interrompent brutalement, autant dans l'art, les techniques, que dans les populations. Plus rien, depuis la Loire jusqu'à Vladivostok, ne rappelle cette intrusion aussi éphémère que limitée. Mais les contacts se rétablissent avec l'art dit « Levantin », c'est-à-dire limité à la côte méridionale de l'Ibérie, où, à nouveau et comme en Sicile, l'animation, la narration, l'anecdote, prennent la place du mythe plastique, exactement comme en Afrique contemporaine. Les liens furent donc maintenus mais via la côte méridionale, exactement comme ce fut le cas au détroit sicilo-tunisien: leur terre était notre mer. Ces contacts, si intenses et si fructueux, ne se limitèrent désormais plus aux marges des océans, car tout le mouvement, dénommé « cardial », en Europe, suit exactement la néolithisation maghrébine, beaucoup plus ancienne, et dont la technologie lithique (pression) et la céramique (incisée) suivent simplement les traditions multimillénaires des récipients végétaux propres à l'Afrique du nord, où les récipients céramiques s'imposaient d'autant moins que les courges y furent abondantes et qu'elles ne s'accommodaient guère aux néolithiques pastoraux africains (fig. 5). Pour le Chalcolithique, un processus analogue s'observe. Les

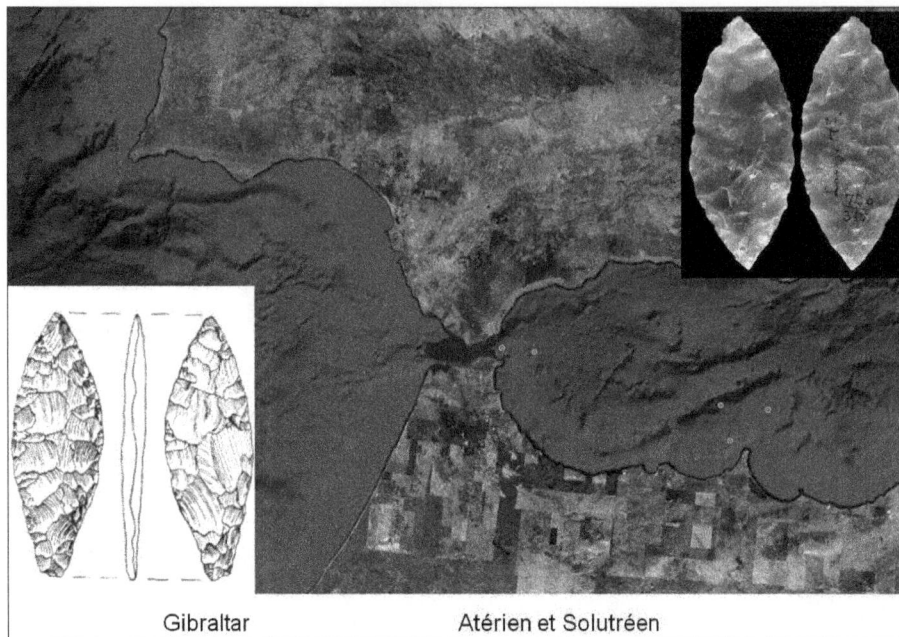

Figure 4. L'Atérien final (28.000 ans) de l'Atérien magrébin, là où précisément l'assèchement accentuait les mouvements de contacts. Seule la phase bifaciale et foliacée concerne cette migration (plus ancienne en Espagne qu'en France) mais les différentes formes de « Proto-Solutréen », typiquement européen, doivent être rattachées aux traditions du Gravettien final

Figure 5. Les éleveurs africains, passant des récipients globulaires dérivés des courges vers la forme céramique et aux décors structurellement analogues au cours de ce passage marin, pour donner la civilisation cardiale

vases dits « campaniformes » en Europe dérivent clairement des vanneries cintrées africaines dont ils héritent les décors cannelés horizontaux, et jusqu'aux croix du fond d'où partent les tressages (fig. 6). En Europe, les deux courants littoraux, atlantiques et méditerranéens, s'amorcent tous deux à partir du point de contact avec l'Afrique, précisément à Gibraltar d'où les premiers métaux natifs furent dispersés.

Le détroit de Bab-el-Mandeb, au sud de la Mer Rouge, fut traversé régulièrement dès l'Acheuléen, présent en égale importance des deux côtés. Les contacts terrestres furent douteux, mais les côtes furent très proches à certaines périodes (fig. 7). Quoiqu'il en soit, les passages furent réalisés avec constance et très anciennement. Les décalages chronologiques soulignent à nouveau une antériorité africaine, logiquement soutenue par le

*Figure 6. Le Campaniforme européen évolua selon les mêmes traces par les deux côtes,
atlantique et méditerranéenne en figeant les formes dérivées du panier vers celles, incompréhensibles
autrement, des vases galbés européens, et apporte les métaux selon les cours des fleuves*

*Figure 7. La diffusion de l'Acheuléen vers l'est de l'Afrique, franchit le détroit de Bab-el-Mandeb,
au sud de la Mer Rouge. Le passage pouvait à la fois être très court et franchissable à l'aide
d'embarcations des plus rudimentaires*

déséquilibre démographique en faveur de l'ouest dans ce cas. Les études récentes sur l'intensité d'occupation de la plaque arabique durant le paléolithique moyen laissent supposer un mouvement inverse à cette époque tous les vestiges furent découverts en abondance du côté oriental du détroit (Crassard et Thiébaut, 2011).

À Ormuz, les contacts furent encore plus évidents: le golf persique exondé, les passages terrestres furent permanents (fig. 8). Et on peut suivre, au Zagros, la même évolution technique qu'en Afrique orientale, de l'Oldowayen à travers tout l'Acheuléen (Otte *et al.*, 2004). Ainsi, la colonisation des Indes paraît très naturelle mais ce mouvement culturel s'interrompt là, sans présager des mouvements ultérieurs et, surtout, antérieurs, quelles qu'en furent les directions.

Le détroit de Wallace fut le plus intriguant: 150 km de haute mer durent être traversés, et ils le furent effectivement dès 60.000 ans au moins, entre l'Indonésie

Figure 8. Au détroit d'Ormuz, le golf arabique fut totalement exondé par intermittence.
Dès l'Oldowayen et durant tout l'Acheuléen, les passages furent aisés
entre la Péninsule Arabique, le Zagros et les Indes

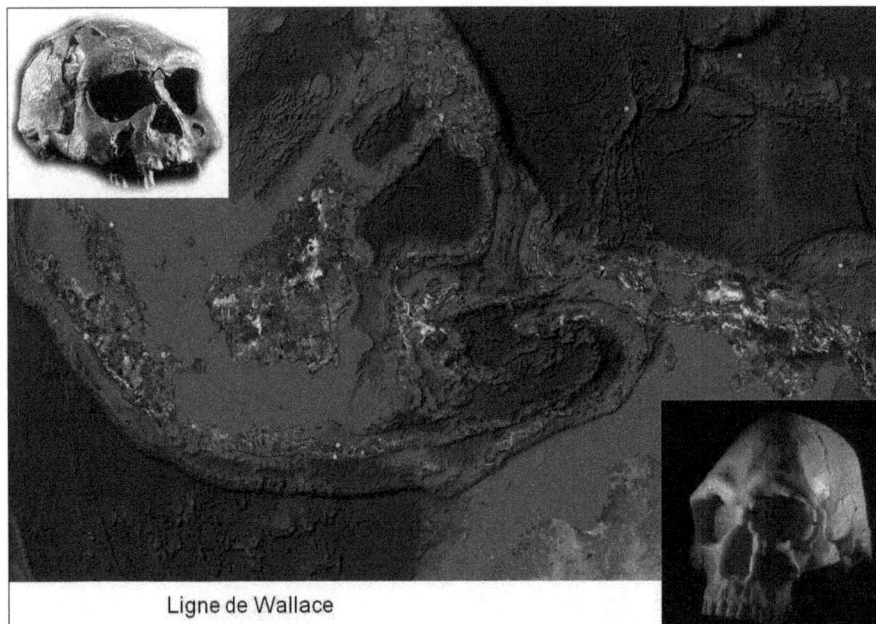

Figure 9. Les passages les plus spectaculaires traversent la ligne de Wallace où 150 km
de haute mer séparaient l'Indonésie de la plateforme de Nouvelle-Guinée. Or,
ces passages de haute mer furent réalisés au moins dès 60.000 ans

orientale et le continent qui rassemblait alors l'Australie à la Nouvelle-Guinée. Populations, mythologies et techniques s'équivalent encore aujourd'hui de chaque côté de cet immense bras de mer (fig. 9, 10). L'un de mes amis navigateur m'évoquait comment, en mer, on « sent » la terre bien avant toute vision directe. Les courants changent, les ciels se transforment, les nuages, les odeurs, la faune, la flore changent (oiseaux, poissons) les couleurs de l'eau, ses mouvements. Tout marin expérimenté sait où il va, quelles sont les distances à parcourir et les directions à suivre, en particulier via la disposition des étoiles, carte céleste infaillible.

Que dire alors des immensités maritimes traversées lors des conquêtes des îles polynésiennes, toutes occupées et installées dans des réseaux de relations régulières, cycliques, à l'inverse du hasard qu'aurait laissé guider d'innocents navigateurs un peu dérangés. Les familles furent embarquées, avec vivres et boissons, la colonisation s'est faite avec régularité, détermination et lucidité. Or, à l'inverse des détroits, ici les terres à joindre ne forment que de minuscules points dispersés sur des immensités océaniques. Il a fallu repérer, revenir, embarquer et retrouver ces points minuscules perdus au milieu des flots. Quelles qu'en soit les solutions

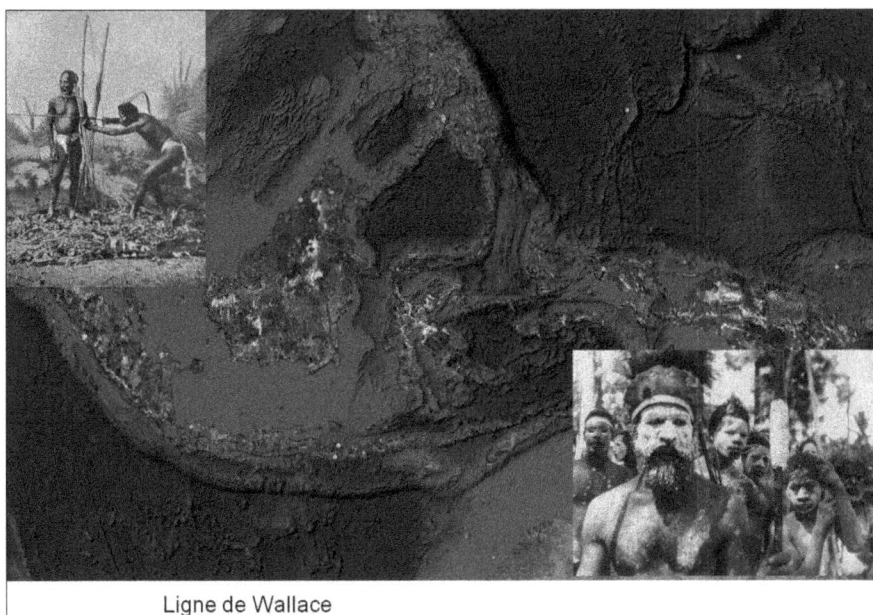

Ligne de Wallace

Figure 10. Les coutumes, les pratiques et les techniques actuelles témoignent des perpétuels échanges entre ces populations liées par la mer

Béringie

Figure 11. Dès l'abaissement des eaux marines, la plateforme continentale, à faible profondeur, reliait la Sibérie à l'Alaska. Par ailleurs, les populations actuelles vivent encore par et pour la mer, le long des côtes actuelles et des archipels dont les îles se joignaient en basses eaux

finalement mises au point dans de telles entreprises, elles ont fonctionné durant des millénaires, portées sans doute par une invincible volonté de conquête, aussi irrationnelle que puissante. Celle qui nous fit grimper sur la Lune, sans autre but que d'y être allé.

La Béringie correspond au cas le plus net de ces traversées intenses, répétées en masse et, sans doute, dans les deux sens. Une fois exondée, la plateforme continentale fut facilement franchie en périodes froides où la mer descendait nettement plus bas que ses hauts-fonds (fig. 11). Mais la masse glaciaire et les hautes montagnes barraient ensuite la route migratoire vers l'Est qui ne présentait d'ailleurs aucun intérêt. Le soleil brillait au sud, avec sa chaleur et ses promesses réconfortantes, c'est donc par là que les migrations successives se sont précipitées, le long des côtes californiennes, alors exondées. Ces migrations resteront à jamais inaccessibles, tant qu'une recherche sous-marine adéquate ne sera pas mise au point (sur le modèle de la grotte Cosquer en Europe).

En outre, et de toutes façons, des deux côtés du Pacifique nord, les populations vivent toujours par voie marine,

*Figure 12. La variation des niveaux marins allonge les plages
et font jaillirent les îles subaquatiques*

autant chez les Aïnous, les Sibériens que les innombrables peuples de la Colombie Britannique. Aucune raison ne permet de supposer qu'il ne fut pas de même au paléolithique le plus ancien. La mer ne paraît redoutable qu'à ceux qui n'en vivent pas: d'innombrables sociétés en ont fait leur habitat, leur mode de transport et la source de nourriture. Les dates très anciennes actuellement disponibles pour les sites brésiliens (40.000 ans au moins) prouvent que ces migrations furent à la fois précoces, intenses et étendues sur d'immenses latitudes, dans un continent totalement déshumanisé et essentiellement disposé selon un axe nord-sud. Les migrations devaient y être rapides dans un désert humain, comme elles le furent dans tous les cas où de fortes différences techniques et mythologiques semblaient « autoriser » de tels massacres humains comme en commirent les colons européens dès la « Renaissance ».

Les détroits fonctionnent donc comme des filtres, tels des estuaires ou de grands fleuves terrestres, sans évoquer les chaînes montagneuses, les déserts, les marécages et les forêts denses. Or, tous ces obstacles furent vaincus et traversés, à diverses reprises et dans tous les sens (fig. 12). Il s'agissait donc d'intensions plus que de moyens pour surmonter ces obstacles, processus bien propres à l'homme depuis qu'il est homme ou qu'il le mérite. L'attrait vers l'impossible est toujours le plus fort, les moyens adéquats s'élaborent ensuite le cas échéant. Les preuves indiscutables (roches taillées, restes humains,

sépultures, foyers, gibiers) sont disponibles dans chaque situation. Il ne s'agit plus de les « attendre » mais de les expliquer, en rendant un peu de leur dignité aux hommes du passé comme nous avons eu tant de mal à la rendre à ceux d'ailleurs.

Bibliographie

CRASSARD, Rémy; THIÉBAUT, Céline (2011) – Levallois points production from eastern Yemen and some comparisons with assemblages from East-Africa, Europe and the Levant. In: Le Tensorer Jean-Marie, Jagher Reto, Otte Marcel (eds.), *The lower and middle palaeolithic in the middle east and neighbouring regions*, Études et Recherches Archéologiques de l'Université de Liège 126, Liège, p. 131-142.

LEROI-GOURHAN, André (1965) – *Préhistoire de l'art occidental*, Mazenod, Paris.

OTTE, M.; BIGLARI, F.; ALIPOUR, S.; NADERI, R.; HOSSEINI, J. (2004) – Earliest human occupations in Central Asia: an Iranian look. In: Derevienko, A. P., Nokrina, T. Ì., *Archéologie et Paléolithique en Eurasie*, Novosibirsk, (en russe), p. 279-282.

PIPERNO, Marcello (dir.) (1999) – *Notarchirico. Un sito del Pleistocene medio iniziale nel bacino di Venosa*, Edizioni Osanna, Venosa.

EXPLORING SIZE AND SHAPE VARIATIONS IN LATE HOLOCENE PROJECTILE POINTS FROM NORTHERN AND SOUTHERN COASTS OF MAGELLAN STRAIT (SOUTH AMERICA)

Judith CHARLIN, Karen BORRAZZO, Marcelo CARDILLO

CONICET-IMHICIHU and University of Buenos Aires. Saavedra 15, 5th floor (1083 ACA), Buenos Aires, Argentina

judith_charlin@yahoo.com.ar; kborrazzo@yahoo.com.ar; marcelo.cardillo@gmail.com

Abstract: *This chapter focuses on the study of variations in late Holocene projectile points from northern and southern coasts of Magellan Strait (Patagonia, South America) in order to explore the existence of cultural divergence processes after the strait formation (9000 BP). We assess size and shape variations, as well as morphological spatial patterns through geometric morphometric and spatial analyses. Our study shows that despite the short geographic distance between northern and southern samples, there are significant differences in projectile point shape, patterns of module covariation and allometry, which exhibit a strong spatial correlation, showing two major morphological clusters on the northern and southern coasts.*

Key-words: *Projectile Points, Fuego-Patagonia, Geometric Morphometrics*

Résumé: *Dans ce travail nous estimons la variation morphométrique des pointes de projectile lithiques durant l'Holocène récent, récupérées au nord et au sud du Détroit de Magellan (Patagonia, Amérique du Sud) afin d'examiner l'existence de processus de divergence (Borrero 1989-1990) après de sa formation. Les techniques morphométriques géométriques et les analyses spatiales ont été utilisées pour comparer les changements des pointes de projectile. Notre étude montre que malgré les courtes distances géographiques entre les échantillons, il y a des différences dans la silhouette des pointes et dans les schémas de covariation et allométrie, montrant deux grands groupes dans la côte nord et sud.*

Mots-clés: *pointes de projectile, Patagonie – Terre de Feu, techniques morphométriques géométriques*

INTRODUCTION

The earliest evidences for human presence in the southern tip of America dated to ca. 10-12,000 yr BP (Bird 1988; Massone, 2004; Nami, 1985-1986, 1987; Prieto, 1991). By that time, the island of Tierra del Fuego was connected to the continent by a land bridge (Figure 1). These first inhabitants were terrestrial hunter-gatherers that shared a common background, both genetic (González-José *et al.*, 2004) and technological. The latter assertion is based on the presence of the projectile point technology known as "Fishtail" (Bird, 1969). By 9,000 yr BP the land bridge was definitively flooded, forming the Strait of Magellan and the Tierra del Fuego Island (McCulloch *et al.*, 1997). Thus, terrestrial populations previously inhabiting the region were isolated by this marine channel. Based on this information, Borrero (1989-1990) first formulated the Hypothesis of Cultural Divergence which states that after this biogeographic barrier appeared, a process of cultural divergence started in Patagonia.

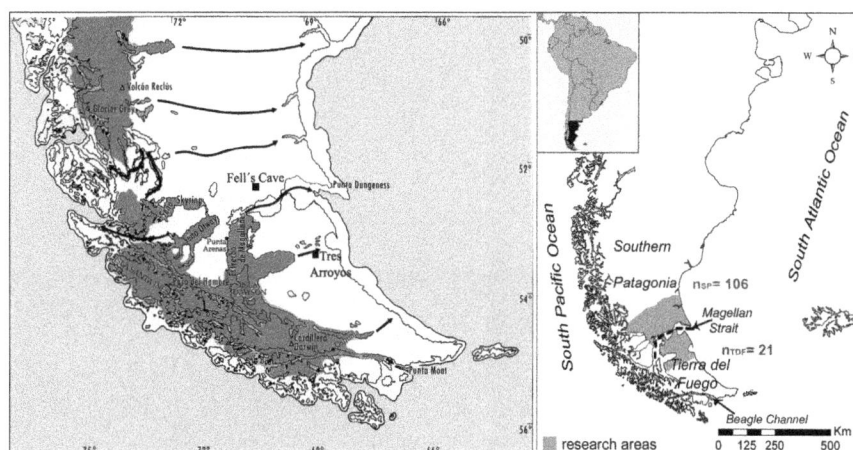

Figure 1. Map of the study region. Left: Southern Fuego-Patagonia by 9,000 yr BP (modified after McCulloch et al., 1997). Right: Research areas (shaded) and sample size

This paper focuses on the study of morphometric variations in late Holocene projectile points recovered from northern (southern continental Patagonia, SP) and southern (northern Tierra del Fuego, TDF) sides of the Strait of Magellan in order to explore the existence of divergence processes in projectile point technology (Figure 1).

REGIONAL BACKGROUND

Systematic archaeological research in SP began in the 1930s with Junius Bird's investigations, who proposed a regional settlement sequence from *ca.* 11,000 BP to historic times (Bird, 1938, 1946, 1988). According to size and shape changes of projectile points, along with other cultural evidences, Bird defined five prehistoric periods (known as Magallanes, Fell, or Bird I to V) previous to European contact. These "cultural periods" have been the main frame of reference for understanding southernmost Patagonia cultural evolution, although subsequent researches provided new data that questioned to some extent this cultural and temporal sequence (see Charlin and González-José, 2012 for a comprehensive review).

The earliest period (Magallanes I) is primarily characterized by the presence of Fishtail projectile points (Figure 2). It was dated to 10,720 ± 300 and 11,000 ± 700 yr BP in Fell's Cave (Bird, 1988). Later on, Fishtail projectile point was also identified at the Tres Arroyos 1 site (Jackson, 2001), which is the only late Pleistocene archaeological assemblage in TDF (10,200 and 10,500 yr BP; Massone, 2004). Only three Fishtail points were

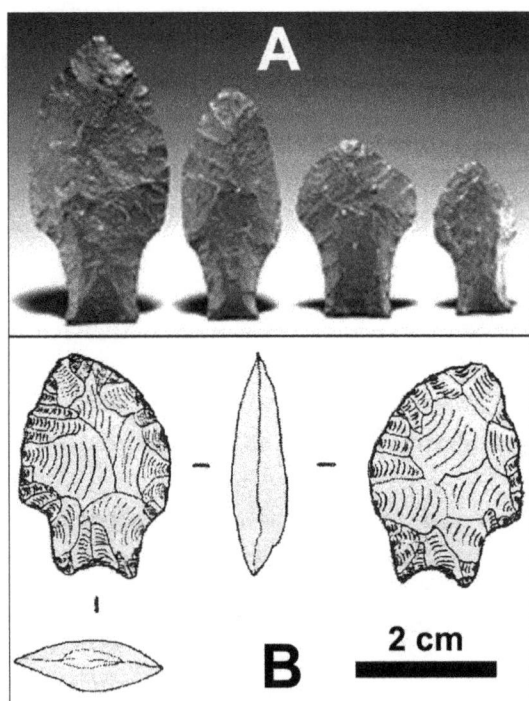

Figure 2. A) Fishtail projectile points from Fell cave, Magallanes, Chile (Junius Bird's collection, courtesy César Méndez). B) Unique entire Fishtail preform recovered from Tres Arroyos 1 (after Jackson, 2001)

recovered in the early compound of this site, which are the only Fuegian samples available for this projectile technology (Figure 2).

The intermediate periods (II and III) were defined on the occurrence of bone and triangular stemless projectile points, respectively (Bird, 1946, 1960).

Periods IV (ca. 3,500 BP) and V (700 BP) correspond to late Holocene and were identified by the presence of Patagónicas and Ona points, respectively (Bird, 1946. Figure 3). Bird called type V points as "Ona" because these continental artifacts were similar to those used by Selk´nam or Ona, an ethnographic terrestrial hunter-gatherer group inhabiting TDF (Borrero, 2001; Chapman, 1986[1982]).

Even though subsequent finding questioned the diachrony between IV and V point types (Gómez Otero, 1986-1987, 1987; Massone, 1979, 1989-1990; Nami, 1984; Sanguinetti de Bórmida, 1984), this typological classification is still in use.

A recent work focused on the reliability of the discrimination between IV and V point types controlling artifact changes after resharpening have shown that these types can still be distinguished in terms of whole point size and stem shape (Charlin and González-José, 2012).

In the case of TDF, the similarity of late Pleistocene and late Holocene projectile points with those corresponding to Fell I and IV and V on the mainland, along with the absence of a comparable chronological schema, prompted the sporadic application of the sequence proposed by Bird in Fuegian archaeology. However, there has not been a systematic study to assess the accuracy of this extrapolation (Huidobro, 2010).

Although Bird's primary research had focused on the mainland, he made a brief characterization of southern island archaeology through the study of cultural materials recovered from shell middens on the Beagle Channel. Thus, he distinguished two periods (Early and Recent or Yaghan) (Bird, 1943, 1946). The Yaghan period includes a projectile point type which subsequent research proved was present all over the island (Borrero, 1979, Figure 4).

RECENT RESEARCH ON PROJECTILE POINT TECHNOLOGY IN FUEGO-PATAGONIA

Several studies have addressed projectile point variations focused on technological (Álvarez, 2009; Franco *et al.*, 2005, 2009, 2010; Gómez Otero, 1987; Gómez Otero *et al.*, 2009; Huidobro, 2010; Nami, 1984), morphometric (Cardillo and Charlin, 2010, 2012; Castiñeira *et al.*, 2009, 2011, 2012; Charlin and González-José, 2012) and performance analyses (Gómez Otero *et al.*, 2011; Ratto, 1990, 1991, 1992, 1994).

Despite the numerous investigations on this technology, only few cases include the direct comparison between SP

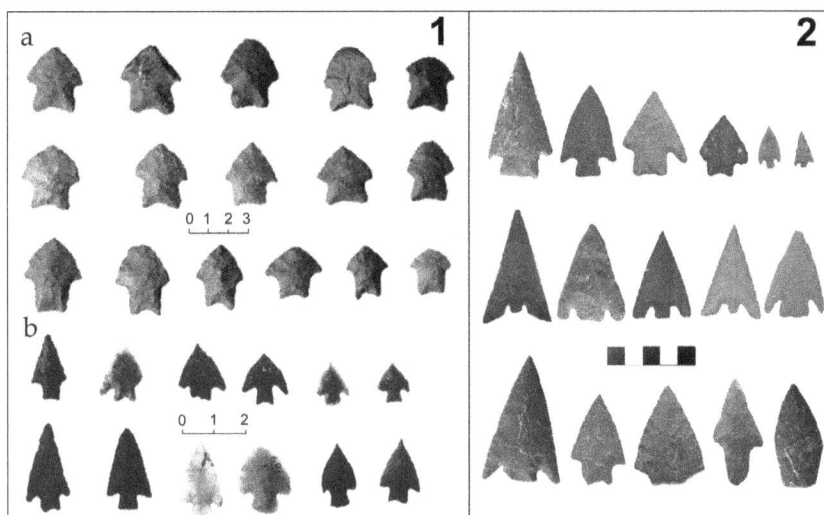

Figure 3. 1) Fell-, Bird- or Magallanes-IV (a) and V (b) points, or Patagónicas and Ona points, respectively (after Bird, 1988). 2) Late Holocene projectile points from northern Tierra del Fuego

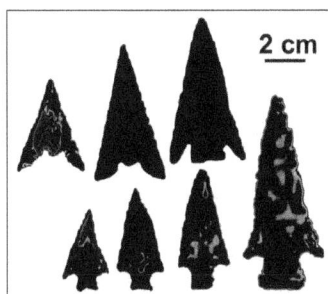

Figure 4. Yaghan projectile points type (Tierra del Fuego) (modified after Bird, 1943: pl. 12)

and TDF. Ratto (1990, 1991, 1992, 1994) developed the first systematic functional studies to assess the optimal performance of southern Fuego-Patagonia late Holocene lithic points through an analysis of design variables. She distinguishes three main technical systems in order of abundance: throwing spear, bow and arrow, and thrusting spear. In the case of TDF, a fourth functional point type was identified, a hafted hand tool, probably used to fleshing sea mammals (Ratto, 1991).

Castiñeira *et al.* (2009, 2011, 2012) and Cardillo and Charlin (2010, 2012) performed geometric morphometric and cladistic studies on Fishtail and Late Holocene projectile points, respectively, from Patagonia, Pampa (Argentina) and Uruguay that highlighted a clinal pattern of variation.

STUDY AREA

Most of the study area located north of Magellan Strait between 51° 26' and 52° 16' S is occupied by a volcanic field known as Pali Aike (PAVF, D'Orazio *et al.*, 2000) (Figure 1). This region includes low and high density sites, being the latest far more usual between the Chico River and the Magellan Strait, where the longest

occupation sequences were recorded (Barberena, 2008; Borrero and Charlin, 2010; Charlin, 2009). Although the human presence was dated *ca.* 11,000 BP in the PAVF Chilean portion (Bird, 1938, 1946, 1988), the broader region was effectively occupied by ca. 4,000 BP.

Tierra del Fuego Island is located on 52-54° S, 66-74° W (Figure 1). Its northern portion is covered by grassland vegetation characteristic of the Magellanic steppe (Collantes *et al.*, 1999). The record of Fuegian occupations has a strong late Holocene signature, although mid-Holocene sites are not exceptional (Salemme and Bujalesky, 2000; Salemme *et al.*, 2007; Morello *et al.*, 2009). Archaeological sites occur in diverse environmental settings: marine cliff, rockshelters, sand and clay dunes, among others. In the study region, larger and intensely reoccupied sites are more common on marine coastal environment, while inland sites are usually smaller and ephemeral.

PROJECTILE POINT SAMPLES

Late Holocene projectile points considered for this analysis were recovered from areas located up to 100 km from Magellan Strait coasts (Figure 1). The sample includes 127 digital images of complete projectile points (SP=106 and TDF=21). They were obtained from authors' research and published literature.

METHODS

Data acquisition

First step in data acquisition was to compile the digital images. The presence of a graphic scale on them was needed to estimate the size of each piece. Then, geometric morphometric techniques were applied to obtain shape and size data from images. These methods allow describing size and shape changes of an object based on Cartesian coordinates in two dimensions using reference points called landmarks and semi-landmarks (Bookstein 1982, 1996-1997).

Generalized Procrustes Analysis was used to remove the effects of translation, rotation, and scaling of the images through the superimposition of point configurations (Rohlf and Slice, 1990). This step generates a matrix of pure shape coordinates.

The size of projectile points is estimated using the centroid size (Dryden and Mardia, 1998). Both sets of data behave independently in the absence of allometry and can be used to characterize and compare artifacts assemblages from their outlines.

The raw images were compiled and scaled in the Tps programs (Rohlf, 2008a, 2008b) and a total of seven landmarks and seventeen semi-landmarks were digitized on the outline of each projectile point. Then, a Procrustes fit was obtained (Rohlf, 2007), and shape and size matrix were exported to the MorphoJ program (Klingenberg, 2008) to perform the statistical analyses.

Data analysis

The analyses focus on three dimensions of data: shape and size of projectile points, and the relationship between sample spatial distribution and morphological similarity. Hence, several statistical analyses were used. The main trends in morphological change were explored with a Principal Component Analysis (PCA) performed on the covariance matrix of shape coordinates. The Discriminant Function Analysis (DFA) was used to compare the mean shape of whole projectile point between research areas as well as between two selected techno-functional modules: the blade and the stem. The covariation between these two blocks of shape variables was assessed by means of a Partial Least Squares Analysis (PLS) (see Klingenberg, 2008; Zelditch *et al.*, 2004 for details).

Mean size of projectile points in SP and TDF samples was compared using a T-test on the centroid size data. The variation in size on shape relationship (allometry) was evaluated through a Regression Analysis. Regression free-of-size residues were also used in a DFA to explore the shape discriminating power while allometric variation was controlled.

The correlation between shape and spatial distribution was tested by Spatial Eigenvector Mapping (SEVM) with a Moran's I autocorrelation measure on longitude and latitude coordinates for each specimen. The method decomposes spatial data at different scales. These new vectors are used as filters (spatial variables) in statistical analyses. These procedures were done with SAM program (Rangel *et al.*, 2010).

Finally, the sample size difference effect on the assumptions of each statistical analysis was controlled.

RESULTS

Shape and Size Analyses

Figure 5 depicts the PCA first two principal components, which explain 86% of total shape variation in the supra-regional scale. The outlines in the graph show the main shape changes. Lighter grids point out relative shape expansion whereas darker grids depict contraction.

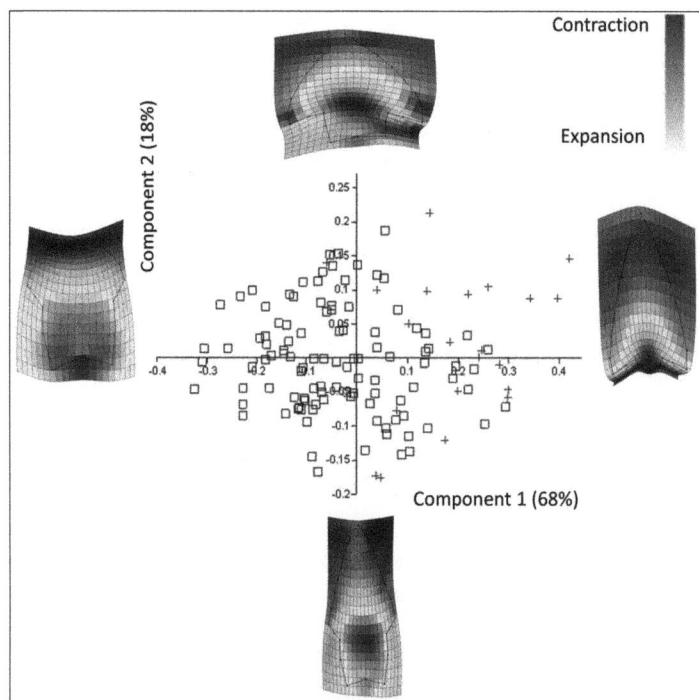

Figure 5. Principal Component Analysis on shape projectile points matrix. Squares: SP. Crosses: TDF

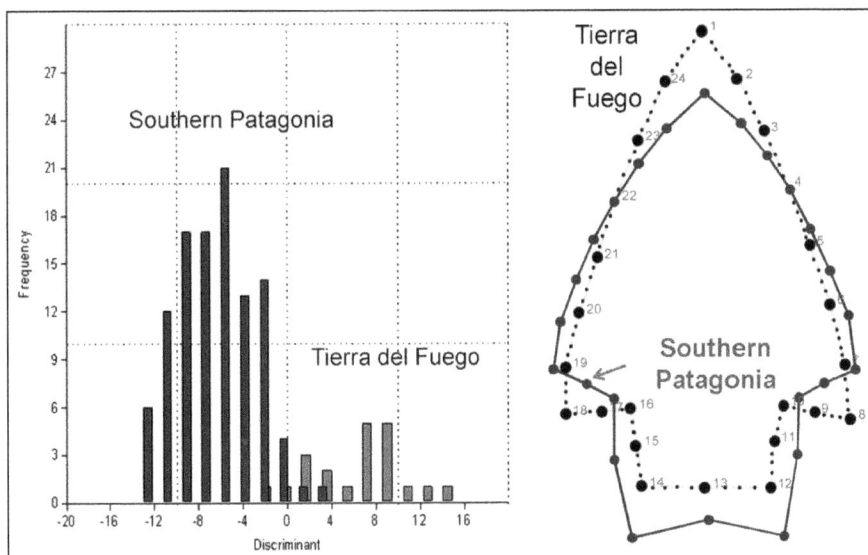

Figure 6. Mean shape differences between SP and TDF projectile points

*Table 1. Blade and stem shape Discriminant Function Analysis
(classification/misclassification tables)*

Blade shape				Stem shape			
From discriminant function: True allocated to				From discriminant function: True allocated to			
Group	TDF	SP	Total	Group	TDF	SP	Total
TDF	20	1	21	TDF	19	2	21
SP	5	101	106	SP	14	92	106
From cross-validation: True allocated to				From cross-validation: True allocated to			
Group	TDF	SP	Total	Group	TDF	SP	Total
TDF	16	5	21	TDF	18	3	21
SP	13	93	106	SP	14	92	106

The first principal component (PC1), which explains 68% of total variance, depicts changes in the relative expansion and contraction of both blade and stem: from elongated blades and contracted stems on the positive scores (right in Figure 5) to the inverse pattern on the negative scores (left in Figure 5). This PC also shows changes in shoulder angle, from acute to obtuse ones (from right to left).

The PC2, explaining 18% of variance, shows relative wide blades and small stems in the positive scores (top of Figure 5) and the opposite in the negative scores (bottom of Figure 5). As can be observed, while SP samples are grouped to the left, TDF samples are clustered to the right, depicting different shapes. These shape differences were tested using a DFA performed on shape coordinates. As PCA suggested, significant differences between SP and TDF mean shapes (Figure 6) were obtained through multivariate T-square test (Hotelling's t^2 238.79 p= <0.001).

In order to test if variations affected stem and blade differentially, we performed a new DFA on two separated shape modules: blade (T-square: 209.0431, p<0.0001) and stem (T-square: 82.3582, p<0.0001). The results indicate that the differences maintain for the two modules between both samples (Table 1).

By means of PLS, we assessed the covariation between the two shape modules (Figure 7). A Permutation test against the null hypothesis of independence (10,000 randomizations rounds) between the modules shows significant results for SP (RV coefficient=0.0847, p=0.002), while it does not for TDF (RV coefficient=0.0894, p=0.519). It suggests a weak blade and stem covariation pattern in SP which contrasts with the module independence recorded in TDF. Furthermore, while blade is the module with major shape variation in SP, stem is the most variable one in TDF (Figure 7).

Differences between group mean size and variance were statistically assessed using a Student t-test (centroid size data transformed to logarithm). The results show there are no differences between TDF and SP projectile points (F=1.42 p=0.36; t=0.85, p=0.34).

13

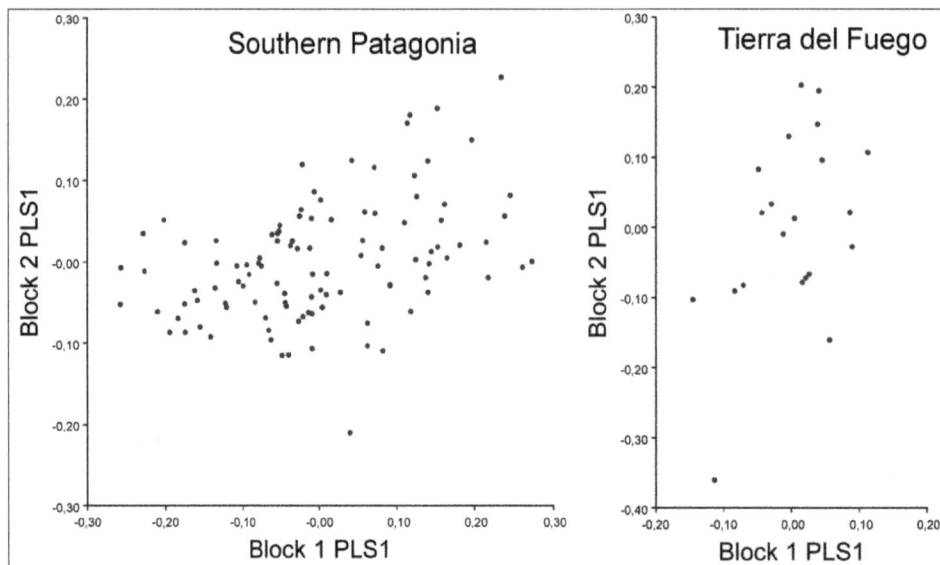

Figure 7. Patterns of shape module covariation in SP and TDF. Block 1: Blade Block 2: Stem

Despite the absence of differences by centroid size, we test the relationship between shape and size within TDF and SP projectile points by performing a Regression Analysis. The results show size and shape were significant correlated within both samples. However, this relationship intensity varies: while the size change predicts 40% of TDF shape variation, it only predicts 3% in SP sample (Table 2).

Table 2. Results of the Regression Analysis between size and shape

Sums of Squares	TDF	SP
Total SS	0.56	2.77
Predicted SS	0.22	0.09
Residual SS	0.34	2.68
% predicted	40	3.4
p-value	0	0.0213

A DFA was performed on the residuals of the shape on size regression. The results keep showing significant differences in free-of-size shape coordinates between SP and TDF (Hotelling's t^2=238.79, p= <0.001) and improve the percentage of projectile point correctly classified to 100% (Figure 8).

Spatial Analysis

A Moran's I Eigenvector Analysis was performed to assess the correlation between shape and spatial vectors. Two spatial vectors were obtained from geographical

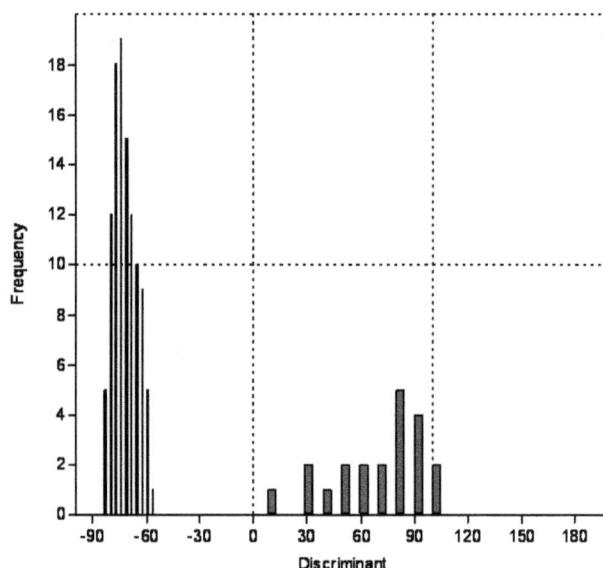

Figure 8. Discriminant Function Analysis on the free-of-size residues

coordinates (latitude and longitude) for each individual. Only the first spatial vector shows significant relation with the shape variation first PC (R^2= 0.27, p<0.001). Results support the idea that morphological variation is spatially constrained on both Magellan Strait sides. In Figure 9, samples with similar colors depict comparable Moran's I autocorrelation values. Moran's I indicates a significant patchiness in mean shape. Thus, similarity among samples described by the first PC variation is related to distance: closer samples are more similar in mean shape than distant ones. Autocorrelation values indicate two different sample groups on each Magellan Strait side (gray scale, Figure 9). In addition, points located near the Atlantic and Pacific coasts (light gray dots) may be suggesting the existence of a third group.

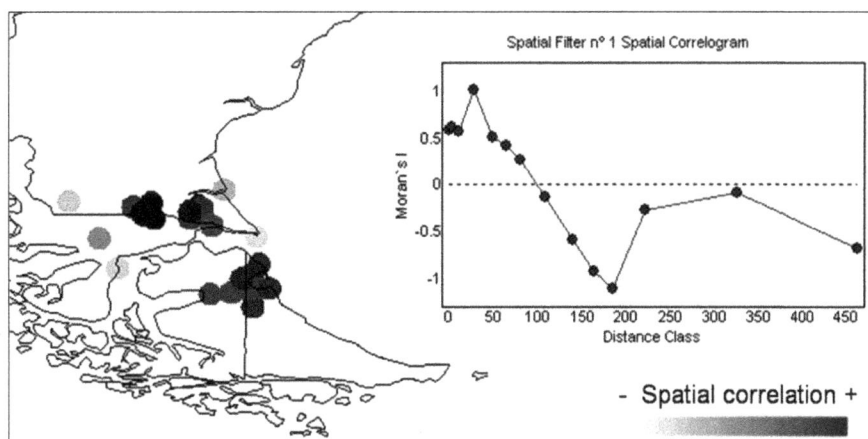

Figure 9. Spatial filter with significant correlation with morphological variation. Darker dots show high spatial autocorrelation (left side). On the right, correlalogram showing abrupt change in spatial structure

DISCUSSION

Several main trends were highlighted by this research. Here, we summarize them and state several hypotheses that need to be test through further analyses.

First, *TDF and SP late Holocene projectile points exhibit differences in the mean shape, considering whole point as well as stem and blade independently*. In addition, *the discrimination between northern and southern samples is improved when using shape residuals (size effect controlled)*. These trends may be suggesting the differential role of functional requirements, design and/or use-life factors in both samples. Since Ratto identified the same weapon systems on both Magellan Strait sides, the recorded differences may respond to variations in the main prey hunted within each region. According to Ratto (1991) as well as stable isotopes data on human remains (Barberena, 2002, 2004; Yesner *et al.*, 1991, 2003) sea food was far more usual in Fuegian diet than in SP hunter-gatherers.

Shape differences can be explained by historical mechanism (neutral or functional *sensu* Dunnell 1980) that in turn, could drive morphological evolution. If historical processes are involved, it is expected that accumulative differences (divergence processes) started with the Strait of Magellan formation and could be traced throughout projectile point lineages from a common ancestor (Fishtails). If not, variation through time must be at random for both areas, as can be expected if variation is related to proximate causes (*sensu* Durham, 1991), like different use-life trajectories and discard rates. Variation can result from a mixture of both factors as well. To test these hypotheses, cladistic analyses are needed.

Second, *there are not size differences between projectile points recovered from both areas*. This may be due to the fact that late Holocene projectile points from SP and TDF were both manufactured on flake or nodules blanks (Borrazzo, 2010; Huidobro, 2010; Nami, 1984, 1986). Most of the lithic raw materials employed within the

study areas are similar size and shape nodules obtained from secondary sources (Borrazzo, 2010; Charlin, 2009; Franco, 1998, 2002; Franco and Borrero, 1999; Gómez Otero, 1985-1986; Nami, 1986).

Third, *allometry is present in both samples but size-related changes in shape exhibit differential intensities, being stronger within TDF projectile points*. The higher allometric change in TDF projectile points may be due to reduction process and use-life extension.

Fourth, *differential patterns of blade-stem covariation were recorded between both areas*. This trend may result from several factors, like use-life trajectories produced from different maintenance and discard rates, and different resharpening strategies between both areas (e.g. in or out the handle).

Finally, *closer samples have similar shapes (spatial correlation). Two main clusters (SP and TDF) were recorded*. The spatial correlation of shape change with the occurrence of the Strait suggests this geographic accident acted as a barrier for projectile point technology of late Holocene terrestrial hunter-gatherers. This trend agrees with others biological and cultural evidences, such as differences in craniofacial morphology (e.g. Cocilovo and Guichón, 1985-86; González-José *et al.*, 2004), size of postcranial skeleton and sexual dimorphism (Béguelin and Barrientos, 2006), presence of rock art (Fiore, 2006) and diversity of lithic assemblages on both Strait of Magellan coasts (Cardillo *et al.*, 2012). Spatial correlation analysis (Figure 9) also indicates the existence of a third group. The spatial provenience of those samples suggests they may be related to maritime environment exploitation. This hypothesis needs to be test on a larger sample.

CONCLUDING REMARKS

This study shows the existence of differences among late Holocene projectile point technology from both sides of the Strait of Magellan. Despite the narrow width of this channel and the short geographic distance between northern and southern samples, abrupt morphological changes occur.

Considering a common late Pleistocene-early Holocene technological substrate, the recorded differences may have resulted from a diversification process that took place after the formation of Strait.

Our study emphasizes the need for a morphometric exploration of late Pleistocene-early Holocene projectile points from Tierra del Fuego and the mainland. Further field work remains as a key issue to enlarge the available samples.

Acknowledgments

We are very grateful to Luis Alberto Borrero, who encouraged this research. His comments helped to improve this paper. This research was funded by Consejo Nacional de Investigaciones Científicas y Técnicas, Agencia Nacional de Promoción Científica y Tecnológica and Universidad de Buenos Aires (Argentina).

References

ÁLVAREZ, M. (2009) – Diversidad tecnológica en el extremo sur de Patagonia: tendencias y continuidades en el diseño y uso de materiales líticos. In Barberena, R.; Borrazzo, K.; Borrero, L.A., eds.- Perspectivas actuales en arqueología argentina. Buenos Aires: CONICET-IMHICIHU, p. 241-267.

BARBERENA, R. (2002) – Los límites del mar. Isótopos estables en Patagonia meridional. Buenos Aires: Sociedad Argentina de Antropología.

BARBERENA, R. (2004) – Arqueología e isótopos estables en Tierra del Fuego. In Borrero, L.A.; Barberena, R., eds.- *Arqueología del Norte de la Isla Grande de Tierra del Fuego*. Buenos Aires: Editorial Dunken, p. 135-165.

BARBERENA, R. (2008) – Arqueología y biogeografía humana en Patagonia Meridional. Buenos Aires: Sociedad Argentina de Antropología.

BÉGUELIN, M.; BARRIENTOS, G. (2006) – Variación morfométrica postcraneal en muestras tardías de restos humanos de Patagonia: una aproximación biogeográfica. Intersecciones en Antropología. Olavarría. 7, p. 49-62.

BIRD, J. (1938) – Antiquity and migrations of the early inhabitans of Patagonia. Geographical Review. XXVIII, p. 250-275.

BIRD, J. (1943) – Excavations in northern Chile. American Museum of Natural History, Anthropological Papers. 38: 4.

BIRD, J. (1946) – The Archaeology of Patagonia. In Steward, J.H., ed.- Handbook of South American Indians. Volume I: The Marginal Tribes. Washington: Smithsonian Institution, Bureau of American Ethnology. p. 17-24.

BIRD, J. (1960) Apéndice I. Period III Stemless Points from Pali Aike and Fell's Cave. In González, A.R., ed.- La estratigrafía de la Gruta de Intihuasi (Prov. de San Luis, R.A.) y sus relaciones con otros sitios precerámicos de Sudamérica Revista del Instituto de Antropología 1. Córdoba: Universidad Nacional de Córdoba. Córdoba. p. 297-298.

BIRD, J. (1969) – A comparison of South Chilean and Equatorial "fishtail" projectile points. The Kroeber Anthropological Society Papers 40, p. 52-71.

BIRD, J. (1988) – Travels and archaeology in South Chile. Iowa: University of Iowa Press.

BOOKSTEIN, F.L. (1982) – Foundation of morphometrics. Annual Review of Ecology and Systematics. 13, p. 451-470.

BOOKSTEIN, F.L. (1996-1997) – Landmarks methods for form without landmarks: morphometrics of group differences in outline shape. Medical Image Analysis. 1:3, p. 225-243.

BORRAZZO, K. (2010) – Arqueología de los esteparios fueguinos. Tafonomía lítica y tecnología en el norte de Tierra del Fuego. Unpublished doctoral dissertation. Facultad de Filosofía y Letras, Universidad de Buenos Aires.

BORRERO, L.A. (1979) – Excavaciones en el alero "Cabeza de León". Isla Grande de Tierra del Fuego. Relaciones de la Sociedad Argentina de Antropología. Buenos Aires. XIII, p. 255-271.

BORRERO, L.A. (1989-1990) – Evolución cultural divergente en la Patagonia Austral. Anales del Instituto de la Patagonia, Serie Cs. Sociales. Punta Arenas. 19, p. 133-140.

BORRERO, L.A. (1994-1995) – Arqueología de la Patagonia. Palimpsesto. Buenos Aires. 4, p. 9-69.

BORRERO, L.A. (2001) – *Los Selk'nam (Onas). Evolución cultural en Tierra del Fuego*. Buenos Aires: Editorial Galerna-Búsqueda de Ayllu.

BORRERO, L.A.; CHARLIN, J. (2010) – Arqueología del Campo Volcánico Pali Aike, Argentina. In Borrero, L.A.; Charlin, J., eds.- Arqueología de Pali Aike y Cabo Vírgenes. Buenos Aires: CONICET-IMHICIHU. p. 1-30.

CARDILLO, M.; CHARLIN, J. (2010) – Diversificación morfológica en las puntas de proyectil de Patagonia. Poster presented at Primer Encuentro de Morfometría "Morfometría geométrica y estudios filogenéticos", La Plata.

CARDILLO, M.; CHARLIN, J. (2012) – Morphological Diversification of Stemmed Projectile Points of Patagonia (Southernmost South America). Assessing Spatial Patterns by Means of Phylogenies and Comparative Methods. In DELSON, E.; SARGIS, E. eds.- Vertebrate Paleobioly and Paleoanthropology Book Series. Springer. In press.

CARDILLO, M.; CHARLIN, J; BORRAZZO, K. (2012)- Artifactual and environmental related variations in

Fuego-Patagonia (Argentina). In Shott, M. ed.- *Lithic technology and the status of rocks in archaeology: a tribute to George Odell*. Utah University Press. Submitted.

CASTIÑEIRA, C.; CARDILLO, M.; CHARLIN, J.; FERNICOLA, J.C.; BAEZA, J. (2009) – Análisis Morfométrico de los cabezales líticos "cola de pescado" del Uruguay. In Palacios, O.; Vázquez, C.; Palacios, T.; Cabanillas, E., eds- Arqueometría Latinoamericana. Buenos Aires: Comisión Nacional de Energía Atómica. II, p. 360-366.

CASTIÑEIRA, C.; CARDILLO, M.; CHARLIN, J.; BAEZA, J. (2011) – Análisis de morfometría geométrica en puntas Cola de Pescado del Uruguay. Latin American Antiquity. 22:3, p. 335-358.

CASTIÑEIRA, C.; CHARLIN, J.; CARDILLO, M.; BAEZA, J. (2012) – Exploring morphometric variations in Fishtail projectile points from Uruguay, Pampa and Patagonia. Current Research in the Pleistocene. In press.

COCILOVO, J.; GUICHÓN, R. (1985-86) – Propuesta para el estudio de las poblaciones aborígenes del extremo austral de Patagonia. Punta Arenas. Anales del Instituto de la Patagonia. 16, p. 111-123.

COLLANTES, M.B.; ANCHORENA, J; CINGOLANI, A.M. (1999) – The steppes of Tierra del Fuego: floristic and growthform patterns controlled by soil fertility and moisture. Plant Ecology. 140, p. 61-75.

CHAPMAN, A. (1986 [1982]) – Los Selk'nam. La vida de los Onas. Buenos Aires: Emecé Editores.

CHARLIN, J. (2009) – Estrategias de aprovisionamiento y utilización de las materias primas líticas en el campo volcánico Pali Aike (Prov. Santa Cruz, Argentina). International Series 1901, BAR Publishing. Oxford: BAR.

CHARLIN, J.; GONZÁLEZ-JOSÉ, R. (2012) – Size and shape variation in Late Holocene projectile points of southern Patagonia. A geometric morphometric study. American Antiquity. 77:2, p. 221-242.

D'ORAZIO, M.; AGOSTINI, S.; MAZZARINI, F.; INNOCENTI, F.; MANETTI, P.; HALLER, M.; LAHSEN, A. (2000) – The Pali Aike volcanic Field, Patagonia: Slab-Window Magmatism near the Tip of South America. Tectonophysics. 321, p. 407-427.

DRYDEN, I., MARDIA, K. (1998) – Statistical shape analysis. Chichester: John Wiley & Sons.

DUNNELL, R.C. (1980) – Evolutionary Theory and Archaeology. *Advances in Archaeological Method and Theory*, 3, P. 35-99.

DURHAM, W.H. (1991) – Coevolution: Genes, Culture and Human Diversity. Standford: Stanford University Press.

FIORE, D. (2006) – Puentes de agua para el arte mobiliar: la distribución espacio-temporal de artefactos óseos decorados en Patagonia meridional y Tierra del Fuego. *Cazadores-Recolectores del Cono Sur. Revista de arqueología. Buenos Aires*. I, p. 137-147.

FRANCO, N.V. (1998) – La utilización de recursos líticos en Magallania. In Borrero, L.A., ed.- Arqueología de la Patagonia Meridional (Proyecto Magallania). Concepción del Uruguay: Ediciones Búsqueda de Ayllu. p. 29-51.

FRANCO, N. (2002) – Estrategias de utilización de recursos líticos en la cuenca superior del Río Santa Cruz. Unpublished doctoral dissertation. Facultad de Filosofía y Letras, Universidad de Buenos Aires.

FRANCO N.V.; BORRERO, L.A. (1999) Metodología de análisis de la estructura regional de recursos líticos. In Aschero, C.; Korstanje, M.; Vuoto, P., eds.- En los tres reinos: prácticas de recolección en el cono sur. San Miguel de Tucumán: Instituto de Arqueología y Museo (FCN e IML – UNT). Ediciones Magna Publicaciones. p. 27-37.

FRANCO, N.; CARDILLO, M.; BORRERO, L.A. (2005) – Una primera aproximación a la variabilidad presente en las puntas denominadas "Bird IV". *Werken. Santiago de Chile*. 6:1, p. 81-95.

FRANCO, N.; CASTRO, A.; CARDILLO, M.; CHARLIN, J. (2009) – La importancia de las variables morfológicas, métricas y de microdesgaste para evaluar las diferencias en diseños de puntas de proyectil bifaciales pedunculadas: un ejemplo del sur de Patagonia continental. Magallania. Punta Arenas. 37:1, p. 99-112.

FRANCO, N.; GÓMEZ OTERO, J.; GURÁIEB G.; GOYE, S.; CIRIGLIANO, N.; BANEGAS, A. (2010) – Variaciones espaciales en diseños de puntas pedunculadas medianas en Patagonia Argentina: Una nueva aproximación. In Bárcena, J.R; Chiavazza H., eds.- *Arqueología Argentina en el Bicentenario de la Revolución de Mayo*, tomo I, cap. 5, (pp. 281-286). Mendoza: Facultad de Filosofía y Letras de la Universidad Nacional de Cuyo, INCIHUSA-CONICET & ANPCyT.

GÓMEZ OTERO, J. (1986-1987) – Investigaciones arqueológicas en el alero Potrok-Aike (Provincia de Santa Cruz). Una revisión de los Períodos IV y V de Bird. Relaciones de la Sociedad Argentina de Antropología. Buenos Aires. XVII:1, p. 173-200.

GÓMEZ OTERO, J. (1987) – Posición estratigráfica particular de puntas de los períodos IV y V de Bird en el alero Potrok-Aike (Santa Cruz). Actas de las *Primeras Jornadas de Arqueología de la Patagonia*. Trelew: Dirección de Cultura de la Provincia del Chubut. p. 125-130.

GÓMEZ OTERO, J.; BANEGAS, A.; GOYE, M.S.; FRANCO, N.V. (2009) – Variabilidad morfológica de puntas de proyectil en la costa centro-septentrional de Patagonia Argentina: primeros estudios y primeras preguntas. In *Octavo Congreso de Historia Social y Política de la Patagonia Argentino-Chilena. Las fuentes en la construcción de una Historia Patagónica* (pp. 110-118). Rawson: Secretaría de Cultura de la Provincia del Chubut.

GÓMEZ OTERO, J.; GOYE, M.S.; BANEGAS, A.; RATTO, N. (2011) – Cabezales líticos y armas en la cuenca del río Gallegos, Patagonia meridional, Argentina. Proceeding of VIII Jornadas de Arqueología de Patagonia. Malargüe, Mendoza.

GONZÁLEZ-JOSÉ, R.; MARTÍNEZ-ABADÍAS, N.; VAN DER MOLEN, S.; GARCÍA-MORO, C.; DAHINTEN, S.; HERNÁNDEZ, M. (2004) – Hipótesis acerca del poblamiento de Tierra del Fuego-Patagonia a partir del análisis genético-poblacional de la variación craneofacial. Magallania. Punta Arenas. 32, p. 79-98.

HUIDOBRO, C. (2010). Métodos de reducción bifacial del norte de Tierra del Fuego durante el Holoceno medio y tardío. Unpublished underdegree dissertation. Universidad de Chile, Santiago.

JACKSON, D. (2001) – Los instrumentos líticos de los primeros cazadores de Tierra del Fuego. Colección Ensayos de Estudio 4. Santiago: Editorial RIL.

KLINGENBERG, C. (2008) – MorphoJ. Faculty of Life Sciences, University of Manchester, UK. Electronic document, http://www.flywings.org.uk/MorphoJ_page.htm, accessed December 7, 2011.

MASSONE, M. (1979) – Panorama etnohistórico y arqueológico de la ocupación Tehuelche y Prototehuelche en la costa del Estrecho de Magallanes. Anales del Instituto de la Patagonia. Punta Arenas. 10, p. 69-107.

MASSONE, M. (1989-1990) – Investigaciones arqueológicas en la Laguna Thomas Gould. Anales del Instituto de la Patagonia. Punta Arenas. 19, p. 87-99.

MASSONE, M. (2004) – Los cazadores después del hielo. Santiago: Ediciones de la Dirección de la Biblioteca de Archivos y Museo.

MCCULLOCH, R.; CLAPPERTON, C.; RABASSA, J.; CURRANT, A. (1997) – The natural Setting. The glacial and Post-Glacial environmental history of Fuego-Patagonia. In Mc Ewan, C.; Borrero, L.A.; Prieto, A., eds.- Patagonia. Londres: British Museum Press. p. 12-31.

MORELLO, F.; BORRERO, L.A.; TORRES, J.; MASSONE, M.; ARROYO, M.; MC.CULLOCH, R.; CALÁS, E.; LUCERO, M.; MARTÍNEZ, I.; BAHAMONDE, G. (2009) – Evaluando el registro arqueológico de Tierra del Fuego durante el Holoceno temprano y medio: lo positivo de los balances negativos. In Salemme, M.; Santiago, F.; Álvarez, M.; Piana, E.; Vázquez, M; Mansur, M., eds.- Arqueología de Patagonia: Una mirada desde el último confín, Ushuaia: Editorial Utopías. p. 1075-1092.

NAMI, H. (1984) – Algunas observaciones sobre la manufactura de las puntas de proyectil de El Volcán. Informes de Investigación 1. Buenos Aires: PREP. p. 85-107.

NAMI, H. (1985-86) – Excavaciones arqueológicas y hallazgo de una punta de proyectil "Fell I" en la "Cueva del Medio" seno de Ultima Esperanza, Chile. Anales del Instituto de la Patagonia. Punta Arenas. 16, p. 103-110.

NAMI, H. (1986) – Experimentos para el estudio de la tecnología bifacial de las ocupaciones tardías en el extremo sur de Patagonia Continental. Informes de Investigación 5. Buenos Aires: PREP. p. 1-120.

NAMI, H. (1987) – Cueva del Medio: Perspectiva para la Patagonia Austral. Anales del Instituto de la Patagonia. Punta Arenas. 17, p. 73-106.

PRIETO, A. (1991) – Cazadores tempranos y tardíos en la cueva Lago Sofía 1. Anales del Instituto de la Patagonia. Punta Arenas. 20, p. 75-100.

RANGEL, T.F.; DINIZ-FILHO, J.A.F; BINI, L.M. (2010) – SAM: a comprehensive application for spatial analysis in macroecology. Ecography. 33, p. 46-50.

RATTO, N. (1990) – Análisis funcional de las puntas de proyectil líticas del sitio Punta María 2 (Tierra del Fuego). Shincal. Catamarca. III, p. 171-177.

RATTO, N. (1991) – Análisis funcional de las puntas de proyectil líticas de sitios del sudeste de la isla grande de Tierra del Fuego. Arqueología. Buenos Aires. 1, p. 151-178.

RATTO, N. (1992) – Técnicas de caza prehistóricas en ambientes de Patagonia (Tierra del Fuego, Argentina). Palimpsesto. Buenos Aires. 1, p. 37-49.

RATTO, N. (1994) – Funcionalidad vs. adscripción cultural: cabezales líticos de la margen norte del estrecho de Magallanes. In Lanata, J.L.; Borrero, L.A., eds.- Arqueología de cazadores-recolectores. Límites, casos y aperturas. Arqueología contemporánea 5. Edición Especial. p. 105-120.

ROHLF, F. J. 2008a tps Utility program version 1.40. Department of Ecology and Evolution, State University. New York: Stony Brook.

ROHLF, F. J. 2008b tpsDig. version 2.12. Department of Ecology and Evolution, State University. , New York: Stony Brook.

ROHLF, F. (2007) – TPSRelw version 1.45. Department of ecology and evolution, State University. New York: Stony Brook.

ROHLF, F.; SLICE, D. (1990) – Extensions of Procrustes method for the optimal superimposition of landmarks. Systematic Zoology. 39, p. 40-59.

SALEMME, M.; BUJALESKY, G. (2000) – Condiciones para el asentamiento humano litoral entre Cabo San Sebastián y Cabo Peñas (Tierra del Fuego) durante el Holoceno Medio. Desde el país de los gigantes. Perspectivas arqueológicas en Patagonia. Río Gallegos: Universidad Nacional de la Patagonia. II, p. 519-531.

SALEMME, M.; BUJALESKY, G.; SANTIAGO, F. (2007) – La Arcillosa 2: la ocupación humana durante el holoceno medio en el Río Chico, Tierra del Fuego, Argentina. In Morello, F.; Martinic, M.; Prieto, A.; Bahamonde, G., eds.- Arqueología de Fuego-Patagonia: levantando piedras, desenterrando huesos... y develando arcanos. Punta Arenas: Ediciones CeQua. p. 723-736.

SANGUINETTI DE BÓRMIDA, A. (1984) – Noticias sobre el sitio "El Volcán", su relación con el poblamiento tardío de las cuencas de los ríos Gallegos y Chico (Provincia de Santa Cruz, Argentina). *Informes de Investigación* 1. Buenos Aires: *PREP.* p. 5-34.

YESNER, D.; FIGUERERO TORRES, M.J.; GUICHÓN, R.; BORRERO, L.A. (1991) – Análisis de los isótopos estables en esqueletos humanos: confirmación de patrones de subsistencia etnográficos para Tierra del Fuego. Shincal. Catamarca. 3, p. 182-191.

YESNER, D.; FIGUERERO TORRES, M.J.; GUICHON, R.A.; BORRERO, L.A. (2003) – Stable isotope analysis of human bone and ethnohistoric subsistence patterns in Tierra del Fuego. Journal of Anthropological Archaeology. 22, p. 279-291.

ZELDITCH, M. L.; SWIDERSKI, D. L.; SHEETS, H. D.; FINK, W. L. (2004) – *Geometric morphometrics for biologists – a primer*. San Diego: Elsevier.

LATE PLEISTOCENE AND EARLY HOLOCENE AT SITIO DO MEIO (SOUTHERN PIAUI, BRAZIL): A REVISION OF STRATIGRAPHY AND COMPARISON WITH PEDRA FURADA

Fabio PARENTI, Giulia AIMOLA
Istituto Italiano di Paleontologia Umana – Roma

Camila ANDRADE
Universidade Federal do Vale do São Francisco – Petrolina

Leidiana MOTA
Erasmus Mundus Master em Quaternário e Pré-história

Abstract: *Sitio do Meio, in southern Piaui, Brazil, is the second rockshelter presenting fully Pleistocenic dates after Pedra Furada. This paper presents a critical revision of the history of excavations (1980-2000), stratigraphy, chronology, and the archaeological content of the site (sector 2). At least 98 stone tools have been identified and described, all of them being older than 12,6 ky BP, i.e. belonging to the Upper Pleistocenic phase of Pedra Furada 3, as defined in the close reference site. The lithic industry of Serra Talhada phase (lower Holocene) is also presented and compared with paleoindian sites of North-eastern and Central Brazil.*

Keywords: *Upper Pleistocene, Lower Holocene, Brazil, lithic industries, excavations, radiocarbon*

Résumé: *Le Sitio do meio, dans le Piaui méridional (Brésil) est le deuxième abri sous roche de la région ayant livré des dates pléistocènes en dehors du site de la Pedra Furada. L'article présente une révision critique des fouilles (1980-2000), la chronostratigraphie et le contenu archéologique du secteur 2. Une centaine d'outils lithiques sont décrits, ils sont plus anciens de 12,5 ka BP et correspondent à la phase Pléistocène Pedra Furada 3 définie dans le site éponyme de référence. On présente aussi l'industrie lithique de la phase Serra Talhada (Holocène ancien) en la comparant aux industries paléoindiennes du Nord-Est et du centre du Brésil.*

Mots-clés: *Pléistocène supérieur, Holocène inférieur, Brésil, industries lithiques, fouilles, radiocarbone*

INTRODUCTION

Since a long time pleistocenic peopling of Americas is the focus of possibly the most copious and inconclusive debate in prehistory. In the last twenty years, the issues on the top spot of this important but yet unsolved chapter of human and natural history range from the genetic background of the ancestral stock(s) which originated the first Americans (Anderson & Gillam 2000, Perego *et al.* 2009), with the associated issue of the number and timing of peopling events (Lanata *et al.* 2008), the palaeogeographic and palaeoclimatic conditions for such events (Pitblado 2011), as well as the "routes" of peopling (Bryan & Ghrun 2003, Dillehay 2008). At the very core of these hypothesis, theories or simply conjectures is – in the ultimate level – the humble nature of artifacts and the stratigraphical context of each site.

Brazilian *lowlands* have been for almost two decades at the centre of action for the Southern continent because of the presence of several sites with radiocarbon dates > 12 ky and the debated nature of the associated cultural material. Since 1980, in South-Eastern Piaui, the franco-brazilian team led by Niéde Guidon conducted many research projects on rock-art chronology (Guidon 1985), "pre-Clovis" archaeological sites (Guidon & Delibrias 1986, Guidon & Arnaud 1991), as well as palaeontological (Guérin & Faure 2008) and palaeoenvironmental data (Chaves *et al.* 2008) (1). The longest chronological sequence in the region is Boqueirão da Pedra Furada (BPF) rock-shelter which has been published in details (Parenti 2001, Santos *et al.* 2003) after an harsh debate (Meltzer *et al.* 1994, Parenti *et al.* 1996). The final Pleistocene layers at BPF (PF 3) show a slight evolution in lithic tool-kit, although a gap in the radiocarbon sequence between 14,300±210 (GIF 6159) and 10,400±180 (GIF 5862) has been recorded. This gap, as shown below, is partially represented in another rockshelter, named Sitio do Meio (SDM).

In this paper we present the available data on past excavations at SDM, obtained on the basis of published data (Guidon & Andreatta 1980, Guidon & Pessis 1993, Pinheiro 2000) and four unpublished dissertations (Aimola 2008, Andrade 2010, Mota 2010, Pinheiro 2007), focusing both on the existence of cultural remains in the final Pleistocene and on the significance of lithic tool-kit in early Holocene, in the context of the prehistory of North-eastern Brazil.

SÍTIO DO MEIO ROCKSHELTER AND THE ARCHAEOLOGICAL EXCAVATIONS

Southern Piaui is dominated by the contrast between the wide pre-Cambrian plain and the Palaeozoic *plateau* of Piaui-Maranhão basin. The bulk of more than 1,300 sites actually recorded in the *Serra da Capivara* National Park and its surroundings are sandstone shelters with rock paintings, mainly dated to early and mid Holocene. As usual in Northern Brazil, the shelters are mostly created by differential erosion of the *cuesta* cliffs, as in the close (1,500 m) reference site of Pedra Furada (Fig. 1, B), incised in a silty layer interbedded in the Silurian sandstone of *Serra Grande* formation. At SDM the available surface is defined by an impressive talus of collapsed sandstone blocks, noticed as soon as the site was discovered (Guidon & Andreatta, 1980). Behind the blocks is the sheltered area (86 m^2), with red-painted panels of *Serra Talhada* tradition on the wall and engravings on some blocks (Cisneiros 2008). During past excavations hundreds of these blocks have been removed, their lenght ranging from 0.4 to 18 m. The shelter is 64 m long, its width ranging from two to 17 meters due to the irregularities of the sandstone wall (Fig. 1, C). The maximum elevation of the roof is about 29 m, which allows the most intense rainstorms penetrate the outer part of the site, as in BPF. The conventional limit between "inner" and "outer" sector of the site adopted in past excavation reports is almost parallel to the drip line (Fig. 1: C). Collapsed blocks were found in both sectors, excavations in the inner part often had to be interrupted because of the presence of such blocks. The deposition of these is the main reason of the filling of rock-shelters in semi-arid lanscapes, as documented also in the case of BPF. In the majority of shelters in Serra da Capivara area, the filling is usually not older than lower Holocene, but in some cases it has been protected from erosion, which was stronger in the final Pleistocene moister conditions.

The interpretation of data from Sitio do Meio is a difficult task, mainly because of the complicated history of its excavations. About 80% of the upper layers of the site have been excavated, but the lowermost Pleistocenic units have been dug only in the central part of the sheltered area.

From 1973 to 2000, six field campaigns have been undergone, all coordinated by Niéde Guidon (Tab. 1), but in collaboration with different field-leaders, in different sectors or – sometimes – in the same sector, but employying different data recording techniques and with differrent scientific goals. The most difficult task the archaeologists faced in this site was the recognition, definition and removing of each – supposed – sedimentary unit. This is mainly because the lateral discontinuity of sandy lenses in rock-shelters, an old nightmare for prehistoric archaeologists (Bordes 1975), but also because of the abundance and hetereogeneity of spallen blocks. For these reasons, the similar numbers observed in the field reports of SDM does not have any stratigraphic meaning and does not imply in any correspondence nor correlation. Therefore, each specific object or stucture

considered in this study has been submitted to a careful verification of its position before its inclusion in the database with a firm chronological attribution.

In order to reconstruct the sequence of excavations, to have a reliable synopsis of the stratigraphy and to clearly identify the Pleistocenic units of past excavations, one of us (GA) conducted a detailed philologic reading of the documentation available in 2008, proceeding as follows: 1) control of the dip of sedimentary units recorded in the sections; 2) synopsis of field notebooks; the most complete for stratigraphic remarks and topographic details are those of 1978 and 1980 campaigns, which served as a basis to reconstruct a stratigraphy with the dip of each excavation campaign.

The site was discovered in 1973; in March 1978, a small trench measuring one by two meter and 1.1 meter deep was opened, stopped by sandstone blocks. In July 1978, excavations were extended up to an area of 5 x 7 m, with the same depth in each square, except on F and G sectors, because of the presence of collapsed blocks. Two radiocarbon dates were obtained from charcoal: 12,200±600 BP (GIF 4628) and 13,900±300 BP (GIF 4927). In 1980, a trench of 3 x 4 m was dug, one meter deep. At the end of this campaign, an assemblage of 79 lithics (30 on siltstone small slabs) was collected and another two dates were obtained (tab. 2): 12,440±230 (GIF 5403) and 14,300±40 (GIF 5399).

Although yielding interesting dates from the Pleistocene-Holocene boundary, the excavations at SDM were interrupted for over a decade, due to the much older chronology observed at BPF. In fact, the quite old Pleistocene dates obtained there in 1978 and 1980 called for larger surface excavations from 1982 onwards. It is not unreasonable to say that from 1984 to 1988, BPF drew almost any effort by Guidon's team, with one of us (FP) coordinating the main campaigns from 1987 to 1988. SDM excavations were resumed in June 1991 in order to try to confirm the very old dates obtained from BPF. In those days, the huge sandstone blocks at the entrance of the shelter still protected the sections of older excavations, the sequence of which was – obviously – used as reference for the new ones. The surviving modern surface was completely surveyed, and the site was subdivided in inner and outer parts and in 5 sectors (Fig. 1: C): sectors 1, 3, 5 remained preserved and new research was undertaken in sectors 2 and 4, the same ones of the earlier excavations. After the removal of larger blocks, in sector 2 (17 x 16 m) and 4 (17 x 8 m) a new metrical grid system was established and excavation was (supposedly) conducted by natural layers (french *décapages*), and each *décapage* was subdivided in excavation phases. Due to the granulometrical homogeneity combined with a lateral discontinuity of the sediment, sometimes this recording system lead to a certain lack of correspondence between the excavations units (*décapage* or phase) and the sedimentary bodies effectively recorded in final sections. In sector 2 (measuring 272 m^2, the larger one), a surface collecting of re-worked remains was undertaken: post-contact ware,

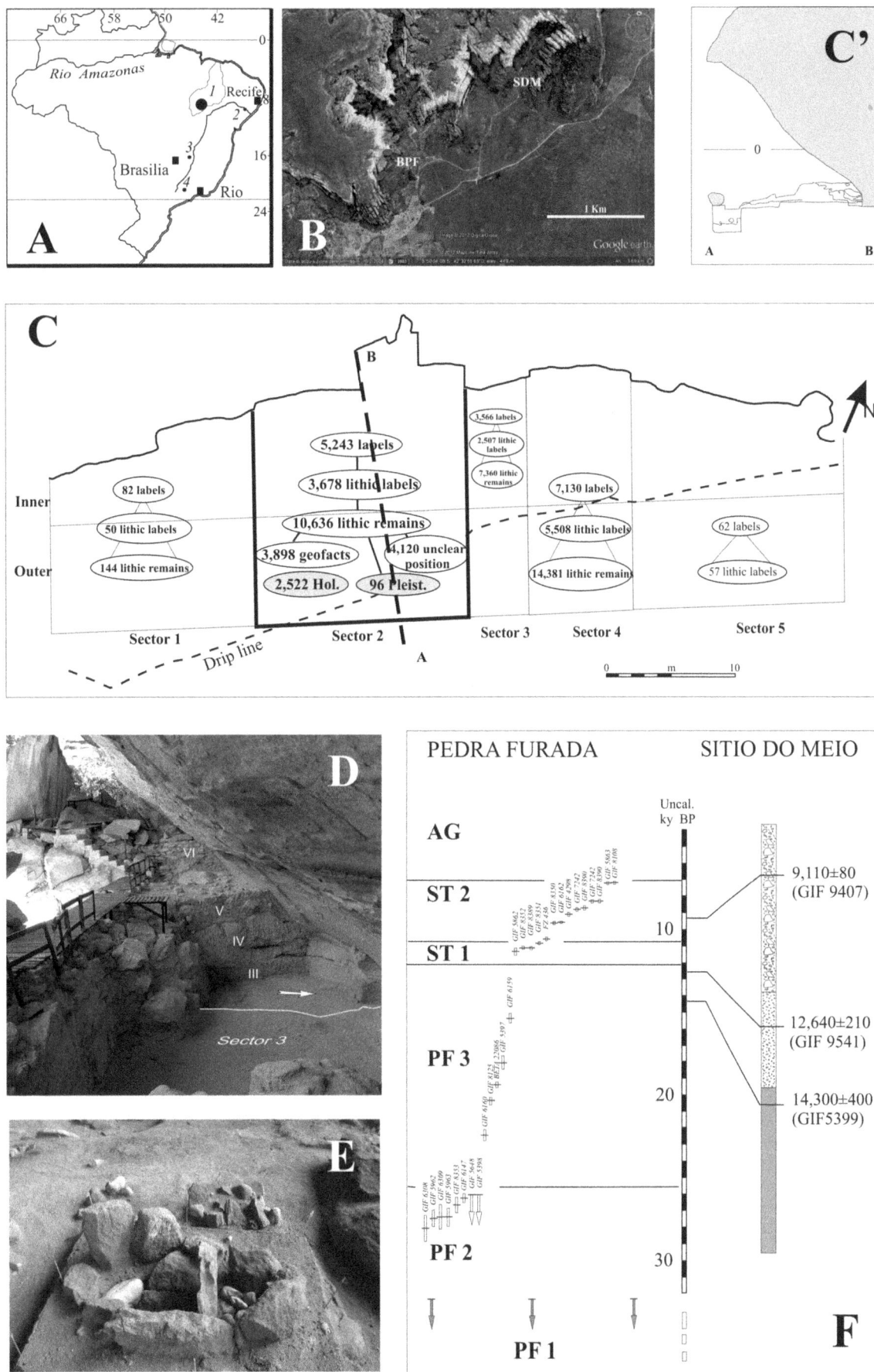

Figure 1. A) Position of quoted sites: [1]SDM, BPF, Perna 1, Paraguaio, Garrincho [2]Sitio do Justino [3]Lapa do Boquete [4]Santana do Riacho; B) aerial view of SDM and BPF; C) site plan with labels and artifacts per sector; C') detail of main section; D) sector 2 and main section, looking West: I, for description, see text, § 2; E) hearth 32, sector 3, 8.804±53 (LY 10138); F) Stratigraphy and chronology of SDM, compared with BPF

Table 1. Excavation campaigns at SDM

Year	Days	Leader	Survey equipment	Dimensions	m³	Progressive m³
1978	13	Guidon	Grid, transit level	5 x 5 x 2.5	62.5	62.5
1980	14	Guidon	Grid, transit level	3 x 5 x 4	60	122.5
1980		Guidon	Grid, transit level	4 x 3 x 2.1	25.4	147.9
1991	47	Team	Grid, transit alidade	2 x 7; 2 x 3	20	167.9
1991		Team	Grid, transit alidade	2 x 2; 5 x 2	14	181.9
1992	12	Equipe	Grid, transit alidade	4.5 x 3 x 5; 9.2 x 2 x 2; 3 x 2 x 5.6	135.5	317.36
1993	6	Equipe	Grid, transit alidade	5 x 4 x 2.4	48	365.36
2000	?	Pinhero	Total station	5 x 5 x 1.6	41.25	406.61
2000	?	Pinhero	Total station	5 x 4 x 1.6	32	438.61
Tot	>92					**2210.64**

lithics, bovids and equids remains, charcoal and modern feces. Below the disturbed sediment, the first true *décapage* was performed: a rich lithic industry, charcoal, ochre and two fragments of painted sandstone were recovered; moreover, a fireplace was identified and excavated. Two trenches for stratigraphic control were opened in this sector. Afterwards the entire sector was excavated and all blocks removed. Although charcoal and lithics were recovered in the lower layers, no datings were obtained in this campaign. In sector 4, two trenches were opened when the excavated surface was at 4th phase of the second *décapage*, in order to attain the same level of 1980 trenches (named – unfortunately – as in sector 2, "Trench 1" and "2", 1991). In 1993, the bottom in sector 4 was finally reached and the outer portion of sector 2 was explored with a trench which was perpendicular to the sandstone wall. Below another collapsed block, more archaeological layers with charcoal fragments and lithics were discovered. On July 29th, the excavations reached the sandstone at the bottom, just below a sterile layer of fine sand, with comminuted (probably natural) charcoal fragments dated at 20,280±450 (Beta 65350) and 25,170±140 (GIF 9542). During the 1993 campaign almost the entire sector 4 was excavated; in its North-eastern part, at 1.3 m from the top-soil, a fireplace was uncovered and dated at 8,800±60 (Beta 47494), and it was not excavated. Despite the loss of some of field notes, we know that one hearth was found, dated at 9.200±60 (BETA 65856) and (apparently) associated with a polished axe (Guidon & Pessis 1993). Excavations were interrupted at an (archaeologically) "sterile" layer at 2.4 m of depth. In 1999 and 2000, P. Pinheiro de Melo carried out new research in sector 4 where the preserved blocks were excavated till the bedrock, providing six fireplaces, the precise position of them, however, is unclear. Sector 3 (6 x 5 m) was also was excavated in order to link the previous sections of sectors 2 and 4; but the term *décapage* was, in this case, referred to natural strata, *i.e.* units of variable thickness, instead of regular excavation phases inside each stratum. After the usual cleaning of modern sediments, 12 fireplaces were unearthed, one of them is morphologically remarkable and dated to 8,805±50 (Fig. 1: E). Just below, another structure with sandstone slabs and an ochre *plaquette*

with a string of graminaceous seeds were recovered along with human teeth from a four to nine years-old individual. This funerary structure was dated at 8,920±50 (LY 10134). The excavation continued untill the *in situ* sandstone, but the available documentation is not very informative and, therefore, it will not discussed here.

STRATIGRAPHY AND CHRONOLOGY

Below the disturbed surface sediments (30-40 cm), the sub-horizontal Holocenic layers are composed mainly by colluvial sands and quartz pebbles with sandstone or siltstone *plaquettes* from the wall. Based on sections available at FUMDHAM files, on notes by J. dos Santos (Santos 2007: 93) and on personal field observations, the sediment can be described as follows, from bottom to top: 1) Palaeozoic sandstone of *Serra Grande* formation; 2) very thin layer of fine sands, thickening close to the outer portion of the shelter; 3) sand with rounded quartzitic cobbles and siltstone fragments; 4) fallen heterometric sandstone blocks; 5) small gravel with angular and sub-angular quartz pebbles in matrix of poorly sorted medium sand. Units 3-5 are visible in Fig. 1: D.

Throughout the 1978-2000 campaigns, a total 29 radiocarbon dates have been obtained, all from charcoal. Among these dates, only 16 (13 in sector 2) have a precise position and will be considered here (Tab. 2). In any case, beneath the 12,640±210 date (GIF 9541), -2.40 m below *datum*, all samples resulted in fully Pleistocenic dates, therefore all the associated material from sector 2 will be considered in this analysis. At the very bottom of the sequence a date of 25,170±140 (GIF 9542) was obtained from a charcoal fragment from the unit 2, a possible fluvial sediment thicker in the outer part of the shelter, not associated with any artifact. This is the *terminus post quem* for the archaeological presence at the site.

LITHIC INDUSTRY

Sector 2 has yelded a gross total of 10,636 inventoried lithic remains (in 3,678 labels fig. 1: C), the vast majority

Table 2. 14C dates from SDM, sector 2 shaded

N°	Unc. BP	+/-	Cal. years	N° Lab.	Field number	Year	Sector	Elevation
1	8.800	60		Beta 47494	39603	1992	2	-1,3
2	8.960	70		Beta 47493	38219/38210	1992	4	-1,64
3	9.110	60		LY 10141	59663/59724	2000	4	-2,12
4	9.110	80	8.342-7.982	GIF 9407	40664/40663	1992	2	-1,60
5	9.200	60		Beta 65856	41542/41540	1993	4	-2,05
6	9.270	100	8.584-8.063	GIF 9408	40758	1992	2	-1,67
7	12.200	600		GIF 4628	2618	1978	2	Liv. V
8	12.440	230		GIF 5403	45	1980	2	Liv. XV
9	12.640	210	13.595-12.289	GIF 9541	40961	1992	2	-2,41
10	12.870	40	13.553-12.941	GIF 9540	40959	1992	2	-2,57
11	13.100	50	13.905-13.338	GIF 9410	40904	1992	2	-2,13
12	13.180	130		LY 6094	40952	1992	2	-2,51
13	13.900	300		GIF 4927	2623	1978	2	Liv. VI
14	14.300	400		GIF 5399	83	1980	2	Liv. XVIII
15	20.280	450		Beta 65350	41302/41304	1993	2	-4,91
16	25.170	140		GIF 9542	41145	1993	2	-5,88 ext.

of which are splitted pebbles of quartz without any intentional flaking evidence (collected either for comparative purposes or because they present evidence of heating) or sandstone slabs that belonged to purpoted fireplaces or other structures. We should note, however, some sampling bias between different field seasons: in 1978 and 1980 only purpoted artifacts had been recovered, whereas in following campaigns all lithic remains were collected. Out of these 10,636 remains, 96 are from Pleistocene and 2,522 from Holocene layers (defined by elevation), 3,851 were classified as geofacts by Pinheiro (2007) and the remaining 4,167 have uncertain stratigraphic position and therefore were excluded from this analysis. Moreover, sector 2 – the largest one – has been extensively excavated and roughly 130 m² are outside the drip-line and, because of this, more exposed to the intrusion of geofacts from the outer stream.

Given this sampling context, the selection of lithics, both from Pleistocene and Holocene units, has been conducted based on the natural gravitative fractures observed in BPF site (Parenti 2001: 135-150). This decision took into account the fact that both, topography and sedimentology at SDM and BPF share similar conditions: both sites lay at the bottom of the *cuesta* and are located between two canyons; SDM, however, is different because: 1) it does not present any active talus of quartz cobbles and pebbles originated by by near waterfalls as BPF (however SDM was probably flooded before the first block collapse); 2) the outline of *cuesta* in front of SDM is convex and narrow, unlike that of BPF, which is large and concave; this probably leads to a lesser water volume during rainfalls. Finally, differently from BPF with its East-West average dip of 10°, sedimentary units at SDM have an sub-horizontal dip; this means that the entry of pebbles

inside the drip-line is much more uncommon than in the outer portion of the shelter, as documented by the fine matrix of sedimentary unit V (Fig. 1: D). In sum, we included in this study any flaked pebble with at least two adjacent flake-scars >10 mm and any flake with striking platform angle < 90°.

The distribution of the 96 confirmed Pleistocenic artifacts of sector 2 is: three choppers, 23 cores, 39 flakes, eight retouched pieces, and 23 fragments. All raw materials are from local origin: quartz and quartzite (66%) and – interestingly – the softer siltstone (32%). The absence of chalcedony and flint in final Pleistocene layers has already been described in BPF and it represents one of the most intriguing archaeological problems in this region (Parenti 2001). No clear hammerstones have been recovered from the Pleistocenic units. Core tools (choppers + cores) presented an average of 3.9 flake scars, (compared to only 2.3 at BPF), pointing, along with a remarkable number of flakes, to a possible function of SDM as flaking stand or – at least in sector 2 – to the presence of a lithic workshop. Flaking is quite expedient also at SDM as in the Pleistocene layers of BPF, because cores have no prepared striking platforms and flakes have mostly (38%) cortical butts. Ten retouched tools have been recorded: three choppers and seven flake tools (Fig. 2).

In the Holocenic units of sector 2, 2,522 lithic artifacts have been recovered, all from the inner portion, in sedimentary units comprised between 9,400±60 (GIF 9027) and 8,100±90 (GIF 9409), subdivided as follows: two hammerstones, 29 choppers (1,1%), 207 cores (8,2%), 579 flakes (23%), 1,549 chunks and small flakes (42%). Quartz and quartzite, locally available, sum up to 95%, with some artifacts made on silicified sandstone

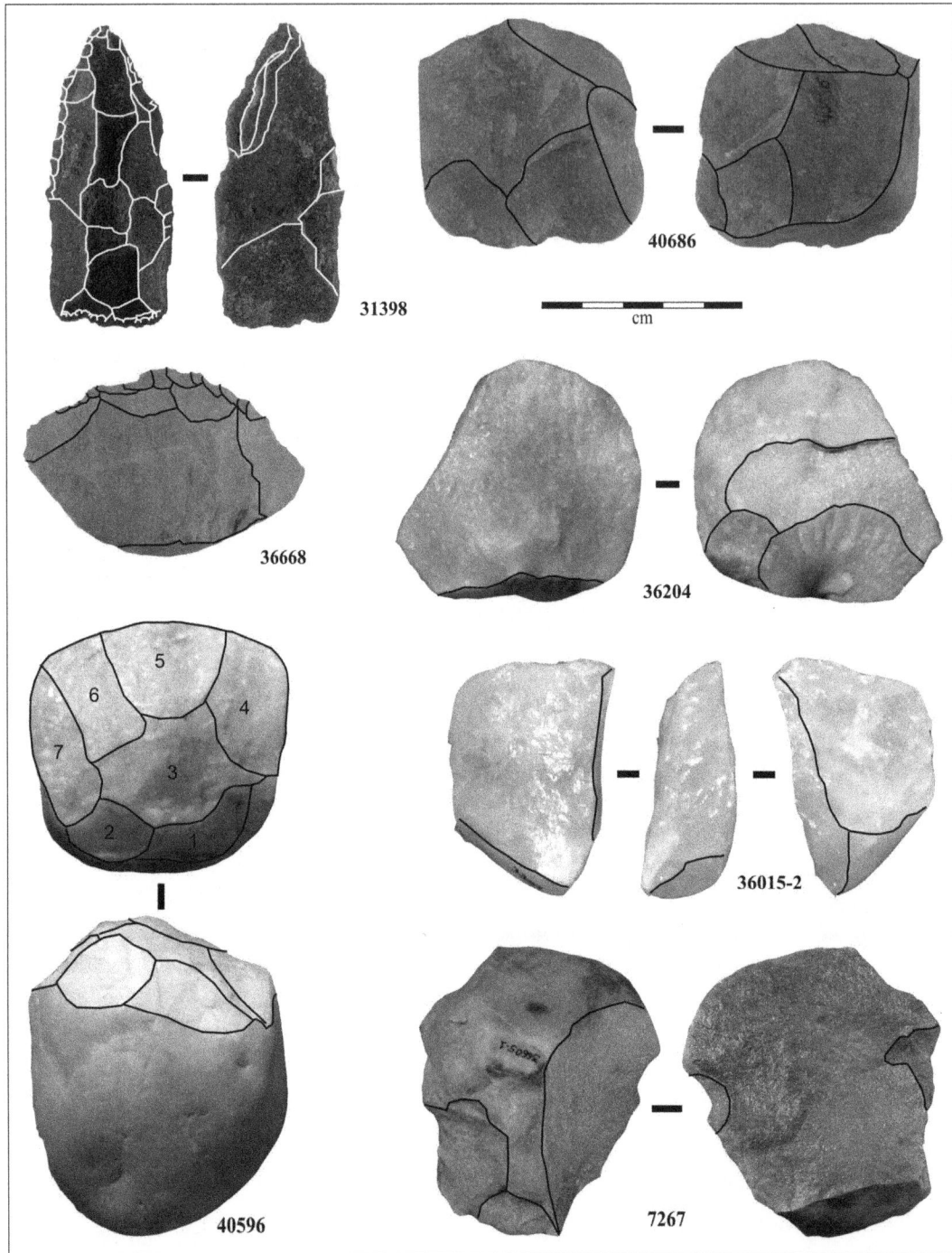

Figure 2. Artifacts from Sitio do Meio, numbers refer to FUMDHAM inventory. Serra Talhada phase (Lower Holocene): 31398) Limace, quartzite; 40686) double scraper, quartzite; 36668) transversal scraper, quartzite. Pedra Furada 3 phase, older than uncal. 12,640 BP: 40596) core on quartz pebble, note the high number of flake-scars; 36015-2) quartz flake; 36204, 7267) quartz flakes, with flake scars on dorsal surface

(1%), chalcedony (2%) and siltstone (2%). The 207 cores present an average weight of about 200 g and they retain 6.5 flake-scars each, almost twice of the observed in the Pleistocene layers. Flakes have mostly cortical butts (61,3%), but plane (36,2%), facetted (0,35%), and punctiform butts (1,6%) were also observed. Eleven flakes have been made by bipolar technique.

In such rich deposits as the ones observed during the lower Holocene in Brazilian Northeast, morphological typology still makes sense, mainly for comparative purposes. For this reason we have organized the 82 SDM Holocene retouched artifacts according to the type-list previously adopted for BPF: 1) half of the artifacts (40) were made from flakes, mainly quartzitic flakes; 2) core tools sum up to 35%; 3) only one *limace* is present; this is an unifacial scraper completely retouched, traditionally considered as the reference tool for the lower Holocene in Brazilian *Planalto* (Lourdeau 2010). It is worth noting that only five tools were made from chalcedony, the prized raw material from which the majority of finished tools have been usually obtained in the rock-shelters toolkit of Southern Piaui. The majority of small flakes have been struck from endogenous rocks, meaning a greater use of endogenous rocks in comparison to the BPF Holocenic layers.

STONE STRUCTURES, PAINTINGS, PALYNOLOGICAL AND FAUNAL REMAINS

A total of 57 structures, including fireplaces, cairns and lithic workshops, have been recorded at SDM, but reliable documentation is available for only 10 stone structures from sector 4 and five structures from sector 2 (all structures from sector 2 were dated). Structures were made mainly of small sandstone slabs, usually fragments that felt from the roof, and quartz cobbles. The majority of such structures has no clear borders and do not show any evident differences in their surface as observed in BPF. However, some of the structures, are impressive, clearly structured hearths (as the one presented in Fig. 2: E) or – sometimes – possible funerary cairns. Remarkably, the dating of one structure (n° 1/18, 1992) gave a final Pleistocene age: 13,180±130 (LY 6094).

A very peculiar and popular artifact from SDM is a kind of anvil, unifacial or bifacial, obtained from large siltstone slabs; 115 of them have been recovered, almost all in sector 2, and 77 attributed to Pleistocenic layers (Pinheiro 2007). Optical analysis using a binocular microscope suggest they have been used to as polishers or grinders for seeds, nuts and pigments, the last ones frequently recovered as red or yellow ochre. In this regard, parietal art at SDM (paintings and engravings) is attributed to Nordeste tradition, dated to lower Holocene (Cisneiros 2008), based only on stylistic ground. Some figures have been engraved on fallen blocks after 8.100±90 (GIF 9409). Finally it is worth mentioning that a polished axe in granodiorite has been described (Guidon & Pessis 1993) as coming from a layer dated at 9.200±60, in sector 4 (Beta 65856) as well as a fragment of coarse ware from sector. 2, dated at 8,960±70 (Beta 47493). Both remains, although relevant, need to be considered with caution because of the complex stratigraphy of the site, as well as the problems associated to the documentation and field recording of the excavations.

Palaeoenvironmental data from SDM come from studies on palynological and faunal remains. Pollen has been obtained from 30 coprolites and the most relevant result (Chaves 1997) shows that, between 12 and 8 ky BP, vegetation was typical of *Cerrado* (arboreal savanna of northern South America), and around 7 ky a *Caatinga* formation (shrubby savanna, as the current one) took over, pointing for a moister climate at the Pleistocene-Holocene transition. Because of the sediment acidity, faunal remains have been recovered only in the upper layers. They consist in about 2,150 remains of small mammals, very similar to the current faunal composition (Von Schmalz 1999).

SITIO DO MEIO IN ITS CONTEXT

At the local level, with regards to stone industry, it seems that SDM did not have, at least based on the portion of the deposit that has been analyzed so far, the central function that BPF had in the organization of a regional settlement system. If the presence of lithic artifacts in the last millennia of Pleistocene at SDM can only be reliably compared with the site of BPF, between 11 and 8 ky BP, an overview at a macro-regional scale can be established. Both SDM and BPF share main technical traits of the Palaeoindian so-called *Itaparica* tradition of Brazilian *Planalto*, present on a wide area, from Goiás to Pernambuco states (Lourdeau 2010). This tradition has been characterized by generally unifacial flaking with occasional occurrence of bifacial projectile points, as observed inLapa do Boquete, Santana do Riacho, Perna I, and BPF. In some cases, like SDM, Santana do Riacho and Sítio do Justino, bipolar flaking has been observed.

In this context the palaeoanthropological evidence, for the period encompassed by SDM sequence, can be summarized as follows: 1) at the very end of the Pleistocene an archaic and robust human morphology is present in the region, as documented by the cranial and dental remains from Garrincho, a limestone cave just 15 km SW from SDM (Peyre *et al*. 1998). The remains have been dated at 12,170±40 (Beta 1366204), dating just after the first preserved occupations at SDM (Guidon *et al*. 2000); 2) at Toca do Paraguaio, another sandstone shelter close to SDM with funerary evidence, two penecontemporaneous burials (dated at 8.8-8.5 ky BP) presented two distinct cranial morphologies, one of Austro-melanesian and African morphology and the other much more similar to present mongoloid forms (Bernardo & Neves 2009). The authors of this study propose that in Lower Holocene two different human groups were present in Northern South America at the same time, probably the product of two different migration waves. This is exactly the period in which all the sites in the region show an impressive change regarding essential cultural traits such as lithic toolkit, rock-art and burials.

CONCLUSIONS

In this paper we presented the revision of the available information on the stratigraphy and archaeology of SDM, presenting analytical data on its lithic industries and showing that a considerable amount of evidence from the main sector can be reliably exploited. A minimum of about one hundred artifacts can be undoubtedly associated to the layers dated from the Late Pleistoce, deserving to be attentively considered in the debate about the earliest peopling of South America.

Sítio do Meio is the second archaeological site in North-eastern Brazil with stone tools associated with structures dated from the Late Pleistocene (BPF was the first one). Despite the anthropogenic origin of Pedra Furada artifacts has been questioned, SDM has better chances to be accepted by the scientific community because of the absence of the most relevant stone breaking agents in this kind of site, i.e. waterfalls. Certainly, we should have more data on the sedimentary history of the deposit and on the taphonomy of lower units, which could still be possible to be done on sector 1. In any case, SDM provides a first step to a systematic intra-site analysis of the Southern fringe of the *Cuesta* of Serra da Capivara.

Aknowledgments

The research and dissertations have been possible thanks to the collaboration of Niéde Guidon and Fundação Museu do Homem Americano (São Raimundo Nonato, Piaui, Brazil). The Italian Ministry of Foreign Affairs has partially funded the campaigns. Thanks to Astolfo Araujo for suggestions and to Mercedes Okumura for the revision of English.

References

AIMOLA, G. (2008) – L'industria Litica di Sitio do Meio (Piaui, Brasile): Ricerca del cambiamento culturale tra Pleistocene e Olocene nel Nord Est del Brasile, unp. MA diss. Univ. of Ferrara.

ANDERSON, D.G.; GILLAM, J.C. (2000) – Paleoindian Colonization of the Americas: Implications from an Examination of Physiography, Demography, and Artifact Distribution. American Antiquity. 65 (1). p. 43-66.

ANDRADE, C.A.S. (2010) – Estruturas de Fogueira dos Sítios Arqueológicos do Parque Nacional Serra da Capivara e Entorno. São Raimundo Nonato. unp. MA diss. Univ. Fed. Vale Sao Francisco.

BERNARDO, D.V.; NEVES, W.A. (2009) – Diversidade morfocraniana dos remanescentes ósseos humanos da Serra da Capivara: implicações para a origem do homem americano. FUMDHAMentos. Revista da Fundação Museu do Homem Americano. 8. p. 87-93.

BORDES, F. (1975) – Sur la notion de sol d'habitat en préhistoire paléolithique. Bulletin de la Société Préhistorique Française. 72. p. 139-143.

BRYAN, A.L.; GRUHN, R. (2003) – Some difficulties in modeling the original peopling of the Americas. Quaternary International. 109-110. p. 175-179.

CHAVES, S.A. [et al.] (2008) – Palinologicals analyses of Quaternary lacustrine sediments from "Lagoa do Quari", NE Brazil (PI). FUMDHAMentos. Revista da Fundação Museu do Homem Americano. 7. p. 64-68.

CISNEIROS, D. (2008) – Similaridades e diferenças nas pinturas rupestres pré-históricas de contorno aberto no Parque Nacional Serra da Capivara – PI. Recife. Unp. PhD Thesis. Univ. Fed. Pernambuco. Recife.

DILLEHAY, T. D. (2008) – Early Population Flows in the Western Hemisphere. A Companion to Latin American History. Holloway T. H. (ed.). p. 10-27.

GUERIN, C.; FAURE, M. (2008) – La biodiversité mammalienne au Pléistocène supérieur – Holocène ancien dans la Région du Parc National Serra da Capivara (SE du Piauí, Brésil). FUMDHAMentos. Revista da Fundação Museu do Homem Americano. 7. p. 80-93.

GUIDON, N. (1985) – A arte pré-histórica da área arqueológica de São Raimundo Nonato: síntese de dez anos de pesquisa. Clio, Revista do Mestrado em História. 2. p. 3-80.

GUIDON, N.; ARNAUD, M.B. (1991) – The chronology of the New World: two faces of one reality. World Archaeology. 23 (2). p. 167-178.

GUIDON, N.; ANDREATTA, M.D. (1980) – O sítio arqueológico Toca do Sítio do Meio (Piaui). Clio, Revista do Mestrado em História. 3. p. 7-29.

GUIDON N. & DELIBRIAS G. (1986) – Carbon-14 dates point to man in the Americas 32,000 years ago. Nature. 6072. p. 769-771.

GUIDON, N.; PESSIS, A.M. (1993) – Recent discoveries on the holocenic levels of Sítio do Meio rock-shelter, Piaui, Brasil. Clio, Revista do Mestrado em História. 9. p. 77-80.

LANATA, J.L. [et al.] (2008) – Demographic conditions necessary to colonize new spaces: the case for early human dispersal in the Americas. World Archaeology. 40 (4). p. 520-537.

LOURDEAU, A. (2010) – Le tecnocomplexe Itaparica. Définition techno-fonctionnelle des industries à pièces façonnées unifacialement à une face planedans le centre et le Nord-est du Brésil pendant la transition Pléistocène-Holocène et l'Holocène ancien. PhD Thesis. Univ. of Paris X.

MELTZER, D.J. [et al.] (1994) – On a Pleistocene human occupation at Pedra Furada, Brazil. Antiquity. 68. p. 695-714.

MOTA, L.A. (2010) – Tecno-tipologia Lítica do Holoceno Inicial (9.450-8.100 anos BP) do Setor 2 do Sítio do Meio – Parque Nacional Serra da Capivara – PI. São Raimundo Nonato. unp. MA diss. Univ. Fed. Vale Sao Francisco.

PARENTI F., (2001) – Le gisement quaternaire de la Pedra Furada (Piaui, Brésil). Stratigraphie, chronologie, evolution culturelle. Paris. Ed. Recherches sur les Civilisations.

PARENTI F. [et al.] (1996) – Pedra Furada in Brazil, and its "presumed" evidence: limitations and potential of the available data. Antiquity. 70. p. 416-421.

PEREGO, U.A. [et al.] (2009) – Distinctive Paleo-Indian Migration Routes from Beringia Marked by Two Rare mtDNA Haplogroups. Current Biology. 19 (1). p. 1-8.

PEYRE, E. [et al.] (1998) – Des restes humains pléistocènes dans la grotte du Garrincho, Piaui, Brésil. Comptes Rendus de l'Académie des Sciences de Paris. 327. p. 355-360.

PINHEIRO, DE MELO P. (2000) – Arqueologia Experimental: os blocos com marcas de uso do Sítio do Meio – PARNA Serra da Capivara – PI-Br. Clio s.s. 1. p. 143-159.

PINHEIRO, DE MELO P. (2007) – A transição do Pleistoceno ao Holoceno no Parque Nacional Serra da Capivara – Piaui – Brasil: uma contribuição ao estudo sobre a antiguidade da presença humana no sudeste do Piaui. Unp. PhD Thesis. Univ. Fed. Pernambuco. Recife.

PITBLADO, B.L. (2011) – A Tale of Two Migrations: Reconciling Recent Biological and Archaeological Evidence for the Pleistocene Peopling of the Americas. Journal of Archaeological Research. 19 (4). p. 327-375.

SANTOS G.M. [*et al.*] (2003) – A revised chronology of the lowest occupation layer of Pedra Furada Rock Shelter, Piaui, Brazil: the Pleistocene peopling of the Americas. Quaternary Science Reviews. 22. p. 2303-2310.

SANTOS, J.C. (2007) – O Quaternário do Parque Nacional Serra da Capivara e entorno, Piaui, Brasil: morfoestratigrafia, sedimentologia, geocronologia e paleoambientes. Unp. PhD Thesis. Univ. Fed. Pernambuco. Recife.

SCHMALZ, K.E. VON (1998) – A Toca do Sítio do Meio: coleção microfaunística. Unp. Report, São Raimundo Nonato, FUMDMHAM.

A CULTURA FEITA MATERIAL: OS INSTRUMENTOS RECORRENTES DOS CERRITOS DO BANHADO DO M'BORORÉ (RIO GRANDE DO SUL – BRASIL) E SUAS POSSÍVEIS INTERPRETAÇÕES

Vanessa Barrios QUINTANA

Mestre pelo Programa de Pós-Graduação em História da PUCRS,
Universidade Federal do Rio Grande – FURG

Abstract: *This paper focuses in relationship between people and things discussing how the culture is transformed of symbolic into something material through human actions. From this, we present the context in which the material culture of Banhado do M'Bororé was found and the interpretations made from the analysis.*

Keywords: *Material Culture, Lytic, Mounds*

Résumé: *Ce document met l'accent sur les relations entre les personnes et les choses dans le forum tels que la culture se transforme de symbolique pour quelque chose de matière à travers des actions de l'homme. De cela, nous présentons le contexte dans lequel la culture matérielle de l'Banhado du M'Bororé a été trouvé et les interprétations faites à partir de l'analyse.*

Móts-clé: *Culture Matérielle, Litique, Monticules*

A CULTURA MATERIAL

Cultura material é o produto material da ação do homem usado pela arqueologia como um meio de nos aproximarmos de populações humanas pretéritas às quais não temos mais acesso. Por isso muitos enfoques arqueológicos tradicionais acreditavam que os objetos possuíam apenas uma natureza passiva, por verem-nos enquanto simples produtos dessa ação. Entretanto, como Hodder (1986) salienta a produção de objetos materiais não pode ser um processo passivo, pois eles representam e agem ativamente na sociedade (Rosa, 2007). Coisas desempenham importantes papéis na formação de pessoas, instituições e culturas, e a forma como pensamos e agimos depende tanto dos objetos com os quais nos cercamos quanto da linguagem que usamos ou das intenções que podemos ter: encontramo-nos através das coisas (Tilley, 2008).

> *Inert matter is transformed by social practices or productive labour into a cultural object, be it a product for immediate consumption, a tool or work of art. (Shanks e Tilley, 1992, p. 130)*

A perspectiva tradicional de conceber objetos como entidades passivas levou a uma atualmente tão criticada dicotomia entre sujeito e objeto, que acabou por colocá-los em esferas sociais separadas. Há sempre um afastamento entre pessoas e coisas, material e ideal, que advém da crença de que a cultura material além de ser passiva, possui propósitos estritamente funcionais e utilitários. Dessa forma, por muito tempo todo o desenvolvimento cultural das sociedades foi pensado a partir da **função** das coisas. Os objetos eram explicados a partir da questão 'para que serve isto?', que após uma série de questionamento que surgem a partir da década de

1980 passa a ser substituída pela questão 'o que isto significa?'.[1] Estas análises funcionalistas levaram a uma arqueologia de dados e cálculos em que os seres humanos pretéritos eram tidos como preocupados apenas com a subsistência e tudo que fosse por eles criado teria apenas um caráter utilitário.

> *The construction of archaeological data reflects the peculiar position of archaeology. Neither humanity nor science, art nor analysis, archaeology combines methods and paradigms of both. Early prehistoric archaeology, in particular, continues to be subject to the paradigms of 'normal science' in a way that is changing in periods from the Neolithic onwards. The world of the hunter-gatherer is seen overwhelmingly as dominated by subsistence imperatives (Pluciennik 2002) and is studied in terms of numbers and counts, the sciences of floral and faunal remains, and the physics of radiocarbon. (Pirie, 2004, p. 678)*

Segundo Rosa (2007) o caráter meramente funcional e utilitário da cultura material é questionado por Hodder (1992) que defende a presença de significados nos artefatos que podem ser atribuídos de diferentes formas e em diferentes relações e contextos. A atenção desta visão utilitária era dispensada majoritariamente aos aspectos físicos e às restrições materiais dos objetos, sendo que seu conteúdo significante, seus elementos simbólicos e ideológicos, eram esquecidos.

Outra crítica à idéia das culturas humanas subordinadas às atividades práticas e com caráter utilitário vem do antropólogo Marshal Sahlins (1976 *apud* Rosa, 2007), que sugere a interpretação simbólica ou significativa da

---

[1] Por exemplo Hodder, 1986; Shanks e Tilley, 1992.

cultura. Para ele o homem não "sobrevive" simplesmente em um mundo material, ele sobrevive de forma específica conforme seus próprios esquemas simbólicos. A funcionalidade das coisas tem finalidade cultural e é definida por este esquema significativo, o que é claro não retira o caráter material das coisas.

Nosso mundo é impregnado pela cultura material da qual, segundo Warnier (1999), não conseguimos escapar por momento algum desde nosso nascimento. Ela possui uma importância fundamental na medida em que se encarrega de transmitir e preservar valores humanos em suas relações sociais. Objetos são parte ativa das relações sociais. Para Rabardel (*apud* Viana, 2005) um objeto pode ser considerado uma estrutura dinâmica uma vez que sua utilização também tem um caráter dinâmico. Seu funcionamento é organizado, o que não impede que possa congregar e adequar novas situações se necessário. Como a cultura material é resultado de um processo produtivo e o indivíduo que confecciona um determinado objeto é sempre um sujeito social, este objeto por ele produzido apresenta duas dimensões: uma privada (própria de cada indivíduo) e outra social. Desta forma a cultura material é uma produção social e socializada, mesmo se trabalho de um único indivíduo (Shanks e Tilley, 1992).

Para Glassie (1999) cultura material é exatamente a **cultura feita material**, uma vez que cultura é apenas um modelo mental, interna, invisível, tornado-se tangível somente através das coisas materiais. Cultura material então combina o visível com o invisível, o tangível com o simbólico. Ela inicia com as coisas, mas não precisa necessariamente terminar nelas, pois, por ser cultural, pode nos transmitir ações e pensamentos, impressos nas cicatrizes deixadas pela atividade humana. Estas cicatrizes formam uma cadeia de informações sobre os objetos, um texto que pode ser lido e descrito durante o processo de análise. Pois como afirma Glassie (1999), mesmo não sabendo o que um objeto significa, nós podemos descrevê-lo e, assim como um texto, ele pode ser quebrado em partes e lido como uma composição, uma vez que a forma como ambos são criados (texto e objeto), através de esforços físicos e mentais, os coloca em conexão.

As atividades das pessoas são construídas e organizadas socialmente ao mesmo tempo em que são representadas simbolicamente na forma de linguagem e objetos materiais (Shanks e Tilley, 1992). Uma vez que não temos mais acesso a linguagem destas pessoas, apenas nos resta tentar ler suas atividades no que chegou até nós, a cultura material. Entretanto, esta leitura não é uma tarefa fácil, pois objetos têm sua própria forma de comunicar-se uma vez que reportam à pensamentos e formulações que resistem a formulação verbal, enquanto tentamos obstinadamente decompô-los em palavras (Glassie, 1999).

A tarefa de transcrever objetos em texto além de ser um empreendimento complicado é também arriscada, pois ao fazê-lo diversos elementos são perdidos. Não posso, por exemplo, alcançar determinadas escolhas do artesão nem

os significados que somente existiram em sua mente. Contudo, essa transcrição é necessária para que a cultura material se torne inteligível, uma vez que é através da narrativa discursiva do arqueólogo que objetos ganham sentido, pois "*Artefacts mean nothing. It is only when they are interpreted through practice that they become invested with meanings.*" (Barrett, 1994 *apud* Holtorf, 2005, 60).

Ao nomear e classificar as coisas construímos relações metafóricas e perdemos muito de seus detalhes, no entanto ganhamos formas de torná-los compreensíveis. Conforme Latour para descrever coisas em palavras, nós as manipulamos conferindo-lhes diversas transformações que resultam no objeto tomando forma, indo do concreto para o menos concreto. E cada transformação que o objeto sofre o torna mais móvel, universal, comparável, padronizado, ao mesmo tempo em que o torna menos particular e detalhado (Pirie, 2003).

Assim, cultura material é um meio através do qual pessoas se comunicam e se expressam. Uma vez que um objeto é visto enquanto um signo, adquire diferentes significados conforme o contexto no qual estiver inserido. Coisas contextualmente estruturadas podem ser lidas da mesma forma que um texto. E assim, a cultura material é transformada em texto para permitir as que as pessoas se comuniquem.

> (...) é a análise contextual de seus usos e significados o que possibilita avaliar a importância dos mesmos não apenas enquanto índices de adaptabilidade mas, também, como meios de satisfação das necessidades práticas do cotidiano e como veículo de transmissão de conteúdos simbólicos e afirmação de identidade pessoal e étnica. (Silva, 2002, p. 120-121)

Quando o único vestígio que nos resta de sociedades remotas é a cultura material, é somente através de sua análise que temos a possibilidade de conhecer essas culturas que não mais existem. Analisando e descrevendo a cultura material, percebemos as mensagens nela inscritas, ou seja, os diversos aspectos que influenciaram em sua gênese. E ao inserir os objetos em seu contexto apreendemos os diferentes papéis que podem ter assumido nas sociedades do passado.

Porém, durante a escavação o arqueólogo elimina um contexto (o contexto arqueológico onde se encontravam as coisas), mas ao escrever ele cria outra relação para as coisas. É este o momento em que os primeiros dados são construídos e as primeiras relações com os objetos acorrem, pois como bem salienta Thomas

> a escavação de um sítio arqueológico deve tornar-se um momento de conversação, negociação, contestação e diálogo entre os participantes, que passariam a produzir dados sobre o passado de forma ativa e participativa. Além disso, as contribuições dadas por cada um dos participantes do trabalho de campo estão vinculadas a um contexto mais amplo, onde suas experiências em estudos anteriores influenciam

no desenvolvimento da pesquisa e por conseguinte no seu resultado final. (Rosa, 2004, p. 24)

O método de escavação e tudo o que acontece em campo influencia na imagem que fazemos do passado (Pirie, 2003). A forma como vemos e percebemos o trabalho de campo guiará a forma como iremos transcrever a cultura material em palavras.

Assim como construções, objetos também são pistas da existência de determinado indivíduo (ou indivíduos) em um determinado espaço. Seguimos cada uma delas e sua leitura nos levará a mais pistas de um passado remoto. Mas estas pistas não são simplesmente descobertas, elas são produzidas pelo arqueólogo a partir de seus pressupostos teóricos e metodológicos – o que não significa que sejam 'forjadas'.

Logo será contada a história de como os vestígios arqueológicos descritos neste artigo ganharam vida. Pois é exatamente isto que arqueólogos fazem: traçam narrativas para contar histórias através das coisas.

Uso minha sabedoria de arqueólogo para criar histórias a partir das coisas que outros deixaram para trás. Transformo coisas em narrativas. Mas, diferente de outros cientistas históricos e sociais, que se comunicam diretamente com as pessoas, o diálogo com a cultura material se dá pela atribuição de sentidos ao próprio objeto. (Hilbert, 2006, p. 99)

No dia 24 de abril de 2004 a equipe do LEPA/UFSM teve o primeiro contato com os sítios arqueológicos do Banhado do M'Bororé. Curiosamente, o fato que mais chamou a atenção da equipe foi a quantidade de material lítico que aflorava dos cerritos, como podemos notar através de trechos do diário de campo:

"Começamos nosso dia conhecendo alguns sítios e pudemos perceber a enorme quantidade de materiais líticos na superfície." (Libiane Cargnin de Lima, diário de campo, 24/04/2004)

"Chegamos a São Borja pela manhã e fizemos uma visita rápida a alguns cerritos, onde o material da superfície é abundante e de boa qualidade." (Vanessa Barrios Quintana, diário de campo, 24/04/2004)

"Chegamos de manhã, e conhecemos alguns cerritos. São muitos e com muito material lítico." (Silvana Zuse, diário de campo, 24/04/2004)

Isto ocorreu, é claro, devido ao anseio de encontrar materiais, em especial as tão desejadas pontas de projétil.

Após as visitas iniciaram-se as escavações dos cerritos escolhidos (Butuy 1 e 2), que antes foram medidos e quadriculados. A primeira camada de grama foi retirada e as espátulas foram dando vida aos primeiros vestígios líticos dos cerritos do M'Bororé. Estas verdadeiras pistas que nos informam sobre povos remotos estiveram sob a terra por centenas, talvez milhares, de anos e quando novamente vem à luz, já chagam carregadas de

significados e pressupostos dados pelos pesquisadores. Elas tiveram uma vida que foi soterrada por terra e grama e agora mais histórias são acrescentadas a essas vidas, atribuídas por uma gama de métodos, técnicas e pressupostos teóricos embutidos nas mentes de quem as escava, pois concordo com Holtorf (2002) que afirma que nós adicionamos histórias às vidas das coisas.

Estas milhares de peças líticas antes mesmo de serem escavadas já haviam sido atribuídas a um grupo construtor de cerritos – e desta forma a estrutura também já estava pré-determinada e carregada de pressupostos –, ligadas e uma 'Tradição' pampeana Umbu e classificadas como 'antigas', vestígios arqueológicos de um povo perdido. Todas estas características foram, porém atribuídas **pelo** pesquisador, "constituídas significativamente – no presente" (Holtorf, 2002, p. 55). A primeira e principal delas é a antiguidade do objeto. Ela é que vai determinar se vale a pena guardar a coisa e estudá-la e será atribuída no momento da descoberta pelo escavador: o que nós acreditávamos que era antigo, se tornou antigo e foi guardado; o que nós acreditávamos que era recente ou que não havia sido transformado e/ou utilizado pelos indivíduos que pretendíamos compreender, virou lixo e foi descartado – senão ali, tempos depois na lixeira do laboratório.

Em seguida o objeto é classificado como um lítico, uma cerâmica, restos alimentares; ligado a uma quadricula, a um poço teste ou trincheira; e colocado assim em sua devida embalagem onde alguns números e letras indicarão sua classificação pré-estabelecida. Chegando ao laboratório suas características talvez se alterem ou mais detalhes sejam atribuídos ou talvez simplesmente ele se torne lixo. Foi exatamente o que aconteceu com a coleção em questão. Após ser lavado, o material recebeu mais detalhes, foi classificado, reclassificado (ou, por exemplo, se percebeu que o que se pensava ser uma lasca era na verdade um fragmento de cerâmica), separado entre lascas e núcleos, se identificou a existência de artefatos.

Cada ação dos pesquisadores, cada interpretação dos fatos passados – apenas interpretações dos fatos e não os fatos em si, pois estes não existem mais –, foi aos poucos acrescentando histórias à vida destas coisas, as quais são relatadas aqui nestas linhas não sem acrescentar ainda mais detalhes. Mais histórias, pois ainda nos restam aspectos empíricos explícitos nas coisas, que nunca deixam de ter sua própria materialidade.

A forma como lidamos com os dados, os métodos de observação, descrição e quantificação dos artefatos, tudo isso influencia na imagem que fazemos do passado (Pirie, 2003). Os objetos só ganham significado através do discurso construído pelo arqueólogo. E o discurso do arqueólogo é construído com os artefatos (Hilbert, 2006).

LENDO AS COISAS

A coleção lítica recuperada durante os trabalhos de campo está composta por mais de 9.000 peças, sendo a grande maioria lascas de tamanhos médios e pequenos, estilhas e micro-lascas. Os núcleos são raros e de pequenas dimensões. Há ainda instrumentos como lascas utilizadas, plano-convexos, pontas de projétil e bolas de boleadeiras.

A matéria-prima utilizada é na imensa maioria o arenito silicificado originado entre os derrames basálticos da Formação Serra Geral que podem ter sido adquiridos nos afloramentos circundantes dos cerritos ou de blocos destacados de afloramentos maiores distribuídos na região. Mas há ainda alguns fragmentos de quartzos e calcedônias, em geral lascas de tamanhos bastante reduzidos.

Os métodos de confecção dos artefatos empregados pelos artesãos pré-históricos foram o lascamento por percussão, o polimento e o picoteamento – estes últimos aplicados à confecção das bolas de boleadeiras.

Fazendo um exercício de reflexão, pensemos nos instrumentos como os vetores de uma série de ações que guardam todos essas ações em si, possibilitando que estas sejam lidas e descritas na forma de texto. A leitura desta peças é fruto de suas vidas presentes, desde o momento da retirada do solo até as interpretações feitas aqui. Porém, ao interpretar as pistas da tecnologia de confecção dos instrumentos, aproximamo-nos de certa forma de suas vidas passadas.

Um tipo de instrumento bastante recorrente na coleção são os chamados **raspadores plano-convexos**, muito comuns em sítios caçadores-coletores. Este tipo de instrumento é recorrente em sítios arqueológicos por todo o mundo e análises mais aprofundadas da tecnologia empregada em sua confecção e de marcas de utilização sugerem que receberiam usos diferenciados.

> *(..) o que comumente é chamado de 'artefatos plano-convexos', na verdade são suportes unifaciais: são matrizes que podem ser organizadas em diferentes instrumentos (ou seja, podem receber diferentes UTF's[2] transformativas ao longo do seu bordo). (Mello, 2006, p. 764)*

Na coleção em questão há uma quantidade significativa de artefatos formais representados pelos instrumentos plano-convexos, que apresentam padronização tanto da matriz quanto do núcleo de onde foi retirado o suporte. Foi possível distinguir três categorias tecnológicas de suporte: 1) suportes com nervura-guia; 2) suportes com superfície central plana; e 3) suportes piramidais. Uma vez que tais suportes foram padronizados, houve uma adequação de suas estruturas volumétricas e três tipos diferentes foram identificados para estes instrumentos: 1) prisma triangular; 2) prisma triangular; e 3) piramidal.

Os suportes utilizados na confecção de tais instrumentos foram lascas de plena debitagem, sem presença de córtex,

predominando lascas cujas retiradas de debitagens anteriores produziram uma superfície plana na parte central da face externa ou lascas cuja parte central é definida por uma aresta longitudinal, sendo que esta última aparece em maioria.

Os instrumentos foram confeccionados sobre lascas pré-determinadas, sendo que todas as qualidades do bloco foram levadas em conta desde o momento da escolha da matéria-prima. Os suportes eram volumosos, proporcionando um maior aproveitamento do gume e possibilitando um maior número de reavivamentos. Os ângulos das bordas dos artefatos indicam que eram utilizados na ação de raspar; portanto, atividades de incisão e corte deveriam ser atribuídas às lascas.

Os instrumentos obedecem a um padrão tecnológico de confecção embora apresentem formatos diferenciados. A matriz foi estruturada o que proporcionou uma total sinergia entre as superfícies – cada retirada influencia na próxima. Foram confeccionados a partir do lascamento direto, com retiradas invadentes. Pequenos retoques foram feitos nas bordas e o reavivamento do gume se dava de uma forma bem característica: lascas grandes e, algumas vezes, relativamente espessas eram retiradas com um forte golpe produzindo, assim, um novo gume que era novamente retocado e utilizado ou em alguns casos, utilizado diretamente. Outra característica marcante é a retirada de lascas contrárias ao plano de percussão que ocorre em praticamente todas as peças. É possível perceber aqui uma atividade largamente associada ao uso dos objetos: a reciclagem visando a manutenção dos instrumentos. A partir disto, vê-se que a estrutura de confecção dos instrumentos é estável, o que pode indicar que o método aplicado em sua produção se inscreve na tradição cultural do grupo, uma vez que

> *(...) os aspectos cognitivos e empíricos constituem a herança técnico-cultural de um grupo, porquanto testemunham a experiência adquirida e sucessivamente transmitida de geração a geração, correspondendo ao saber-fazer, relacionado às operações intuitivas baseadas na experiência pessoal do artesão (Boeda, 1997; Karlin e Julien, 1996 apud Viana, 2006, p. 803).*

A confecção adequada dos instrumentos plano-convexos está ligada à eficiência técnica do artesão, pois, a aplicação de conhecimentos tecnológicos complexos exige, concordando com Viana (2006, p. 829-830) "não somente seleção de matéria-prima adequada, obtida com base em 'escolhas' previamente determinadas, mas também conhecimento e domínio dos métodos e técnicas, que cada concepção exige para a eficácia de sua produção".

Uma leitura mais detalhada de cada peça[3] demonstra este **padrão de confecção** referido está ligado ao aprendizado e a herança cultural do grupo, uma vez que parece claramente ter ocorrido o planejamento prévio dos objetos por parte dos artesãos.

---

[2] UTF é a organização particular das retiradas, cujas conseqüências técnicas agem em sinergia para colocar uma característica técnica remarcável e coerente (Mello, 2006, p. 767). *Nota da autora.*

[3] Para uma análise detalhada da cultura material ver Quintana, 2010.

Após a leitura dos instrumentos é possível concluir que os suportes foram lascas bastante robustas com estruturas volumétricas formatadas – sendo identificados três tipos. As principais ocorrências observadas na coleção são as seguintes:

→ Um padrão tecnológico é percebido com determinadas características presentes em todos os instrumentos.

→ Os negativos na face dorsal anteriores ao lascamento de alguns instrumentos formam superfícies planas.

→ As retiradas de *façonnage* geralmente invadentes e abruptas.

→ Retidas abruptas originaram grandes negativos reflexivos.

Os ângulos dos instrumentos apontam para a atividade de 'raspar', sendo empregado em materiais como peles, madeira e ossos. Entretanto, somente através da análise microscópica dos vestígios de utilização destes instrumentos seria possível afirmar que materiais teriam sido por eles trabalhados. Mas as suposições feitas se baseiam na discussão a respeito dos ângulos dos instrumentos, segundo a qual a ação a ser desempenhada necessita de um determinado valor de ângulo: para ações de raspar o ângulo do gume deve ter em torno de 70° a 90°; para a ação de cortar o gume deve formar um ângulo em torno de 40° a 60°; um ângulo menor que 40° permite cortes deslizantes (Boeda *apud* Viana, 2006, p. 132).

Outro fator que influencia no funcionamento do instrumento é o formato da linha de gume. Linhas de gume curvas são apropriadas para cortar e talhar, sua área de ação é maior e melhor aproveitada. Já as linhas de gume retas são mais adequadas a furar e fatiar, mas limitam-se a uma área de ação menor. Note-se que os instrumentos da coleção em estudo possuem linhas de gume de ambos os tipos, entretanto seus ângulos são maiores que 70°, encaixando-se nas atividades de raspar. Destaca-se ainda que instrumentos com maior ângulo exibem maior resistência, podendo ser aplicados em objetos a serem transformados que exijam maior força motriz do instrumento transformativo.

As pequenas dimensões dos artefatos plano-convexos chamam a atenção, pois apresentam em média um comprimento de 5,65 cm. Estas dimensões reduzidas implicam limitações com relação às dimensões do material no qual seriam empregados os artefatos, mas, em contrapartida, implicam uma maior facilidade no transporte.

Para Kuhn conjuntos artefatuais compostos por uma série de pequenos artefatos unifaciais confeccionados sobre lascas representariam a solução ótima para articular transportabilidade e multifuncionalidade na elaboração dos conjuntos de artefatos transportados pelos caçadores-coletores em diversos tipos de deslocamento, pois apresentam a melhor relação em termos de utilidade e peso (1994 apud Bueno, 2007, p. 88).

Percebemos ainda que as pequenas dimensões dos instrumentos não foram um empecilho para seu reaproveitamento. Um método particular de reavivamento (as grandes retiradas com um forte golpe referidas acima) era empregado com o objetivo de criar novos ângulos e com isso a peça podia ser exaustivamente utilizada.

Devido à grande quantidade de material, sendo a imensa maioria lascas e microlascas sem marcas de uso ou retoques, o foco principal foram os instrumentos recorrentes aqui representados por 18 peças denominadas pela literatura arqueológica plano-convexos. Fazer uma leitura individual mais detalhada de cada objeto permitiu identificar um padrão em sua confecção. O lascamento era direto e com retiradas invadentes; as peças foram exaustivamente retocadas, sendo que ao não haver mais ângulo de percussão outro era produzido através de retiradas rasantes e espessas que geralmente deixaram negativos de lascas reflexivas; foram realizadas retiradas contrárias ao plano de percussão, a partir do ápice da peça, que podem ter servido à uma melhor preensão do instrumento ou para algum tipo de encabamento; rebaixamentos de pequenas porções da superfície ventral dos instrumentos auxiliavam na obtenção de gumes mais agudos e conseqüentemente mais afiados.

As características acima descritas são encontradas em praticamente todos os instrumentos plano-convexos, bem como nas lascas da coleção que embora não remontem aos artefatos aludem a outros confeccionados a partir das mesmas técnicas, mas que não ganharam vida durante nossos trabalhos de campo e que provavelmente continuam a espera de alguém que os ajude a nascer. Dessa forma, os resultados da leitura dos instrumentos plano-convexos permitem ligá-los a um mesmo grupo cultural local.

Análises mais amplas focadas na tecnologia de confecção de artefatos líticos e comparações com outros estudos podem talvez relacioná-los a uma cultura regional, uma vez que artefatos semelhantes são recorrentes em sítios arqueológicos de regiões pampeanas.

Referências Bibliográficas

GLASSIE, Henry (1999) – *Material Culture*. Indianápolis: Indiana University Press.

HILBERT, Klaus (2006) – Qual o compromisso social do arqueólogo brasileiro? In: Revista de Arqueologia. 19: 89-101. (Consult. 05 Nov. 2009) Disponível em http://periodicos.ufpb.br/ojs2/index.php/ra/article/vie wFile/1670/1312

HILBERT, Klaus (2007) – Indústrias Líticas como Vetores de Organização Social ou: um ensaio sobre pedras e pessoas. In: Bueno, Lucas e Isnardis, Andrei. *Das Pedras aos Homens*: tecnologia lítica na arqueologia brasileira. Belo Horizonte: Argvmentvm, p. 95-116.

HOELTZ, Sirlei E. (2005) – *Tecnologia Lítica*: uma proposta de leitura para a compreensão das indústrias do Rio Grande do Sul, Brasil, em tempos remotos. Tese (Doutorado Internacional em Arqueologia) – Pontífice Universidade Católica do Rio Grande do Sul. Porto Alegre.

HOLTORF, Cornelius (2005) – *From Stonehenge to Las Vegas:* archaeology as popular culture. Lanham: Altamira Press.

IRIARTE, José (2003) – *Mid-Holocene Emergent Complexity and Landscape Transformation: the social construction of early formative communities in Uruguay, La Plata Basin.* Tese (Doctored in Philosophy) – College of Arts and Science at the University of Kentucky. Lexington.

MELLO, Paulo Jobim de Campos (2006) – É Possível Perceber Evolução no Material Lítico Lascado? O Exemplo das Indústrias Encontradas no Vale do Rio Manso (MT). In: *Revista Habitus*, vol. 4, n. 2. Goiânia: IGPA/UCG, p. 739-770.

QUINTANA, Vanessa Barrios (2007) – *Manifestações Culturais nas Terras Baixas do Rio Grande do Sul:* os "Cerritos de Índios". 2007. 51 f. Monografia (Graduação em História) – Curso de História, Universidade Federal de Santa Maria, Santa Maria.

QUINTANA, Vanessa Barrios; LIMA, Libiane Cargnin de; MILDER, Saul E.S. (2007) – Manifestações Culturais das Terras Baixas Platinas: os cerritos de índios. In: I Congresso Internacional da SAB, 2007. Florianópolis. *Anais do I Congresso Internacional da SAB [CD-ROM].* Florianópolis: Sociedade de Arqueologia Brasileira. I CD-ROM.

RODET, Maria Jacqueline e ALONSO, Marcio (2007) – Uma Terminologia para Indústria Lítica Brasileira. In: Bueno, Lucas e Isnardis, Andrei. *Das Pedras aos Homens:* tecnologia lítica na arqueologia brasileira. Belo Horizonte: Argvmentvm, p. 141-154.

ROSA, Carolina Aveline Deitos (2007) – *Pessoas, Coisas e um Lugar:* Uma interpretação para a ocupação pré-colonial no sítio arqueológico Morro da Formiga, Taquara, RS. Dissertação de Mestrado. Porto Alegre: PUCRS/FFCH/PPGH.

SCHMITZ, Pedro Inácio; NAUE, Guilherme; BECKER, Ítala Basile (1997) – Os Aterros dos Campos do Sul: a Tradição Vieira. In: Kern, Arno (Org). *Arqueologia Pré-Histórica do Rio Grande do Sul.* Porto Alegre, p. 221-250.

SHANKS, Michael e TILLEY, Christopher (1992) – *Re-Constructing Archaeology.* 2ed. London: Routledge.

SILVA, Fabíola Andréa (2002) – As Tecnologias e Seus Significados. In: *Canindé*, Xingo, n. 2, p. 119-138.

TILLEY, Christopher (2008) – Part I: Theoretical Perspectives. In: David, Bruno and Thomas, Julian (Ed.). *Handbook of Landscape Archaeology.* Walnut Creek: Left Coast, p. 7-11.

TILLEY, Christopher (2008) – Phenomenological Approaches to Landscape Archaeology. In: David, Bruno and Thomas, Julian (Ed.). *Handbook of Landscape Archaeology.* Walnut Creek: Left Coast, p. 271-276.

TILLEY, Christopher (1994) – *Phenomenology of Landscape:* places, paths and monuments. Oxford: Berg Burg Pub Ltda.

TIXIER, J.; INIZAN, M.L.; ROCHE, H. (1980) – *Préhistorie de la Pierre Taillée:* terminologie et technologie. Paris: Cercle de Recherches et d'Études Préhistoriques.

VIANA, Sibeli A. (2006) – *Pré-História do Vale do Rio Manso.* Goiás: UCG.

WARNIER, Jean-Pierre (1999) – *Construir a Cultura Material:* o homem que pensava com seus dedos. Paris, Presses Universitaires de France. Tradução: Emílio Fogaça.

THE ORIGINS OF THE BRAZILIAN SAMBAQUIS (SHELL-MOUNDS): FROM A HISTORICAL PERSPECTIVE[1]

Gustavo Peretti WAGNER

Professor and researcher in postdoctoral stage at the Postgraduate Program in Anthropology,
Federal University of Bahia, Brazil

Abstract: *The theme of the origin of the sambaquis is an interesting one to academic research since the nineteenth century. Attempts to elucidate the question are diverse, and present implications and developments from both human biology and socio-cultural perspectives. The pages that follow seek to address all these implications through a historiographical approach, proposing a discussion of revisional and integrating character, composing an explanatory hypothesis. This paper describes the details of one aspect from the communication presented at the XVI World Congress UIPPS and XVI Congress SAB, in 2011.*

Key-Words: *Sambaquis, possible origins, coastal settlement, historical review*

Résumé: *L'origine des sambaquis est l'une des questions centrales qui ontintéressé la recherche académique depuis le siècle XIX. Les tentativesd'élucidation de cette question sont marquées par une grande variation, selondes implications et des dédoublements, soit du point de vue de la biologiehumaine que du point de vue socio-culturel. Les pages que suivent essayentd'aborder toutes ces implications d'après une approche historiographique, en proposant une discussionqualifiée par la révision et l'intégration, en élaborant une hypothèseexplicative. Ce travail caractérise de façon détaillée un aspect d'unecommunication présentée au XVI Congrès da UISPP et au XVI Congrès de la SAB, en 2011.*

Mots clés: *Sambaquis, origines plausibles, peuplement côtier, révision historique*

INTRODUCTION

Sambaquis are the first archaeological sites considered as such in Brazilian historiography, considering that, since the 16th e 17th century they were identified as remnants of various indigenous activities, such as fishing stations or mollusks collection, burial and rituals (cf. Cardim, 1939[1584]; Madre Deus, 1920[1797]). However, the researches in institutional level began only two centuries later.

The revival of interest for sambaquis can be attested in Rath's (1856) initiatives[2] that when studying the formation of the current coast of São Paulo State mentions the presence of sambaquis in the area, attributing antediluvians dates to it. Decades later, the question of the origin of the sambaquis is central to the archaeological studies, now at the institutional level. This era became known (e.g. Souza, 1991) as the period of Brazilian archeology patronage, having D. Pedro II as the main supporter. His self-interest led him to witness the exhumation of burials in the sambaquis of Sant'ana River, in São Vicente.

That period is marked by the rise of an intense debate about the origins of the sambaquis. Opinions were divided into three streams of thought, as follows: 1) naturalist, represented mainly by Hermann von Ihering, who advocated the natural origin of sambaquis as a result of sea fluctuations and epirogenic movements dating back to the Tertiary, 2) artificialist, mainly represented by LadislauNetto, that considered the sambaquis as a result of prehistoric human activities, and 3) mixed, stream of conciliatory character that became hegemonic from the early decades of the 20th century and consider the existence of large original shellmounds and the presence of archaeological sites formed by artificial accumulations, which could occur overlapping the first (Costa, 1934; Leonardos, 1938; Souza, 1991; Lima 1999-2000).[3]

In the years around the turn of the century the issue mobilized the public opinion, with developments even outside research institutions.[4] In the same period, the participation of experts with different backgrounds such as medicine and geology were multiplied. On the one hand, the approaches used supported the hypotheses raised, but on the other hand, they introduce new theories,

---

[1] Communication presented at the XVI National Congress of SAB and XVI World Congress of UISPP, section 10 – The First Americans and the Origin of Mankind (2), coordinated by Luis Borrero, Laura Miotti and Lucas Bueno.

[2] The study was conducted at Rath's own expenses in 1845. He was an engineer by profession and he was stimulated by personal interest. Years later, Orville Derby undertakes a similar work, producing the outstanding synthesis by Krone (1914) on the sambaquis of the lower valley of the Ribeira de Iguaperiver.

[3] The questions concerning to the geological implicatios of the origins of the sambaquis are present in detail in Wagner (2012, in press).

[4] Koseritz contributed to the issue founding and managing journals in Pelotas and Porto Alegre such as "O Noticiador" (1852-1856), "Gazeta de Porto Alegre" (1875-1888), "Jornal do Comércio" (1868) and "Neue Deutsch Zeltung" (1864-1940) (see Koseritz, 1884). Ihering (1895) refers to the publications in the newspaper in São Paulo in 1889. In fact, individuals connected to the arts ended up stating an opinion on the subject, such as the painter BenedictoCalixto (1904).

leading to the segmentation of the issue, systematized in the following pages as "racial" e sociocultural implications. Since then, the theme has appeared in the archeology of sambaquis, polarizing opinions, the sites being understood sometimes as the result of populations originating from the coast, or as the result of groups from the country inlands that, from a given time, began to explore the productivity of coastal environments.

RACIAL IMPLICATIONS

In the nineteenth century the participation of medical professionals directed approaches to craniometric studies, from which the origins of the coastal settlements was sought through the identification of the "race" responsible for the formation of sambaquis. As a result the opposition between the "race of Lagoa Santa," inhabitant of the interior highlands and the "race of the sambaquis," inhabitant of the coast was created[5] (Lacerda; Peixoto, 1976; Lacerda, 1885).

Lacerda, Peixoto (1876) established the relationship between Botocudos skulls and the skulls found by Lund in Lagoa Santa. The high dolichocephaly observed in the skulls series lead them to propose a link between the "race of Lagoa Santa," the Patagonian and Eskimos, characterizing them as an original race (and indigenous) in America. Lacerda (1885) is responsible for coining the term "man of the sambaquis." The expression is clearly an idea that supports the existence of a racial unity in the Brazilian coast. He characterized it as an invading race that was distributed along the coast, and at a certain point, disappeared. However, he establishes biological relationships between Botocudos from central Brazil and the man of the sambaquis. Thus, the coast occupation is connected to the ancient inland settlement, attributing a "paleo-American" origin to the "man of the sambaquis". Years later Ihering (1904) suggests the existence of two types of people associated with the south and southeast sambaquis, with one associated to the Botocudo type and the other related to the type known then as Tupi, but not exactly the same as the latter.

Imbelloni (1936), (see Sauer, 1944; Willey, 1966), proposes racial unity to the shell mounds sites in South America, with their living descendants of the Botocudo type in the interior of Brazil, uniting them under the fueguid type and laquid type. Years later, from the point of view of a cultural adaptation to the coastal environment, Emperaire; Laming (1958) suggest that the shell mounds sites would constitute a "fringe" of settlement along the entire coast of the South American Continent. In southern Brazil, Mello-Alvim (1978) reinforced this hypothesis advocating biological proximity between the occupants of the sambaquis and the shell mounds of Patagonia, Argentina and southern Chile. Since the decades of 1980-1990 Rivera; Rothhammer (1990) propose a close relationship between the southern coast

of Brazil and northern Chile shell mounds builders when comparing the former occupants of the Camarones 14 to the fishermen and gatherers of the Morro 1 de Cabeçuda.

Two decades later Neves (1988) criticizes the alleged biological uniformity suggested to the shell mounds sites along the Brazilian coast and highlights the need for proper segregation between sambaquis specifically, pre-ceramic shell mounds settlements and ceramic shell mounds settlements. It demonstrates the biological diversity associated with the formation of these different types of sites, but stands in favor of the genetic homogeneity among the builders of sambaquis, noting a movement in population from north to south along the coast, with the introduction of different genetic material in three stages: 1) hunter-gatherer occupation in the central coast of Santa Catarina, 2) ceramist occupation in Itararé in northern Santa Catarina, and 3) Tupiguarani occupation in the same region.

However, it is only in Neves; Okumura (2005) that the hypothesis of an inland origin to sambaquis groups is assumed, settling it in the Ribeira River valley, south of São Paulo. Based on skeletal analysis they show a biological link between the sets from the fluvial sambaquis, the series from south-central coast of São Paulo as well as from the Paraná coast, placing the key to the origin of fluvial sambaquis in the Moraes site.[6]

Okumura (2008) undertook extensive research of cranial morphology in samples from coastal and inland populations in South and Southeast regions of Brazil. When comparing the populations of the interior to the populations of the coast, the results suggested the independent development of both areas. We highlight here the presence of samples from two inland sites: Cerrito Dalpiaz, settlement that defined the Umbu Tradition with chronology between 5950 ± 190 BP 4280 ± 180 BP and Capelinha 1 with burial dates in 8860 ± 60 BP. Masi (2001, p. 113) in a study focused on the central region of Santa Catarina had already shown in detail, through analysis of collagen in individuals from the coast, a clear distinction between the dietary patterns of gatherer coastal fishermen and inland hunter-gatherers, "... who did not migrate to the coast as it was traditionally thought". And Filippini; Eggers (2005-2006) quoted the biological distance between shell mound coastal builders and the shell mound inland builders from sites of São Paulo State.

SOCIO-CULTURAL IMPLICATIONS

What is discussed here under the title socio-cultural implications refers to attempts to clarify the origins of the

[5] Initial synthesis of the issue can be found in Costa (1934), Mattos (1941) and Emperaire; Laming (1956) and, for a recent review, see Okumura (2008).

[6] It is noteworthy that Neves (1988) detected the existence of a "pocket" of genetic material between northern Santa Catarina and northern Paraná and Neves; Okumura (2005) relate the fluvial sambaquis to the sambaquis in south-central São Paulo and Paraná, it expresses the suggestion that the Ribeira Valley represents the original location of the dispersion of sambaquis, at least in the Middle Holocene. It should be noted, however, the existence of higher dates than the sixth millennium in the coast, but the respective chronological regional contexts restrict them. A more detailed discussion on this topic can be found in Wagner (2012, in press).

sambaquis starting exclusively from the material culture contained in sites, the settlement patterns and use of space, the stratigraphic composition and regional established chronologies.

In a letter attached to Wiener (1876), Netto (1976, p. 2) is the first to formulate the hypothesis that sambaquis groups builders have originated from the inland populations of the continent. "...It seems to have been these deposits accumulated during the winter of each year by the tribes of the interior ...". Years later, Netto (1885) located the origin of the sambaquis in the valley of the Rio Paraná, Paraguay, from where migratory waves would have migrated to the valleys of the Amazon and Prata, making a semicircle in eastern South America, reaching the coast. Despite the imaginary nature, the hypothesis predicts the involvement of various indigenous cultures in the construction of the sambaquis (Aymara, Quechua, and other originating from Goiás and Mato Grosso).

Hartt (1885) suggests a Peruvian origin for the Marajó coastal occupations and coast of Salgado and Ihering (1904), linked the Amazon sambaquis to the "developed cultures of the Andes" located between northern Argentina and Mexico. Likewise, he excludes Sambaqui of Porto Santo described by Rathbun (1878) in Itaparica, Bahia, and other sites of the Northeast from the "sambaquiana province", located strictly between São Paulo and Rio Grande do Sul.

Serrano (1937)[7] emphasizes the cultural differences at the regional level, in proposing the division of the Brazilian coast in two cultural facies: southern marked by the presence of zoolithes and northern where zoolithes were nonexistent. In the following decade, he explains the issue and extends the classification grouping the Brazilian sites in four categories: 1 – archaic, formed by Azaraprisca, the oldest, circumscribed to São Paulo and originating from the "culture of Lagoa Santa," 2 – southern, including the sites of Rio Grande do Sul, Santa Catarina, Paraná and south of São Paulo, containing zoólitos from Guayaná and subsequently occupied and/or acculturated by the Guarani, 3 – middle stage, limited to the states of Rio de Janeiro and Espírito Santo, and 4 – amazon (Serrano, 1946). Except for the Amazon, he considers all sambaquis in southern Brazil belonging to laguid e fueguid racial types, according to the classification by Imbelloni (1936), connecting them, thus, to conchales from Tierra del Fuego and the southern archipelago of Chile.

Between the publications of Serrano (1937, 1946), Leonardos (1938) performs a synthesis that covers the entire coast of the country and incorporates all the shell mounds sites from the Amazon coast to Rio Grande do Sul, in the sambaquis category.

In the years before the establishment of the National Archaeological Research Program (PRONAPA),[8] the

studies have been intensified in the South and Southeast, highlighting the work of Bigarella (1950-1951); Emperaire; Laming (1956), Hurt; Blasi (1960), Tiburtius; Bigarella (1960), Rohr (1962), Rauth (1962), Salles-Cunha (1963), to name just a few examples, all sites being brought together under the term sambaqui. In the North, Hilbert (1959) excavated the Sambaqui of Ponta do Jauarí and in the Northeast Calderón (1964) excavated the Sambaqui da PedraOca, demonstrating the wide and free usage of the term.

With the arrival of the National Program for Archaeological Research (PRONAPA) and the diversification of sites dated by C^{14}, it became possible to suggest the original region of the sambaquis on the Brazilian coast. With the incorporation of Rauth to the Program, the research about the sambaquis continued, keeping methodologically marginal in view of the existence of an already structured research orientation, and the sites of Bahia and other states of South and Southeast remained united under the same category.

Since then the archaeological culture of the sambaquis itself has been present in the national syntheses as a phenomenon confined to the south and southeast, following an established tradition of research in the 1950s (Emperaire; Laming, 1956; Schmitz, 1984, 1998; Neves, 1988; Prous, 1992; Lima, 1999-2000; Masi, 2001; Tenório, 2003, 2004; Okumura, 2008; Wagner et al., 2011). Only Simões (1981), Roosevelt (1991), Perota; Botelho (1993) insisted on researching sites in northern Brazil,[9] and the presence of ceramic associated to remote chronologies led to the distinction between the north and south sites of the country, and the boundary was set in the Todosos Santos bay, Bahia State.

Uchôa (2007 [1973], p. 21) stated that it was impossible at that stage of the research, to specify the moment of arrival of the sambaquis builders on the coast, however, "(...) About 6.000 years ago, these people began to move across the Atlantic coast...". A group of researchers recently considered that there is a cultural unity implied to all coastal sites, whether it is regarding the variability of their size, composition of the layers, archaeological structures, material culture or functionality of the site (DeBlasis et al., 1998, Gaspar et al., 2008). Gaspar (1991, 1996) already argued for the existence of a socio-cultural pan-Brazilian unity: "(...) It regards the vestige of a socio-cultural system whose remarkable feature is to associate, in the same space, the location for living,[10] burying the dead and disposing of assets and food leftovers." (Gaspar, 1996, p. 82).

---

[7] Actually, the hypothesis of Serrano (1937, 1946) owe much to the ideas of Ihering (1895, 1904).

[8] Research program of national character funded by the Brazilian National Research Council (CNPq) and the Smithsonian Institution with the participation of 12 Brazilian archaeologists under the general coordination of Betty Meggers and Clifford Evans, between 1965 and 1970.

[9] Researches in North and Northeast regions of the Country are developed in the current work of Silva (2000), Martinelli (2007) and Bandeira (2006).

[10] It should be noted, however, that Gaspar et al. (2008, p. 320) redirected part of this particular issue, stating "(...)It is clear that the sambaquis do not represent ordinary occupations, but instead, they are specialized elements of systems of settlements of which very little is known about other types of sites.(...)"

Lima (1991, 1999-2000) suggests that the origin of the sambaquis would be related to different archaeological cultures or to different socio-cultural systems from the countryside, asserting that "... these piles must be analyzed from the perspective of diversity..." (Lima, 1999-2000, p. 314). In that same study he presents a comprehensive chronological framework and clearly limits the archaeological culture of the sambaquis to the south and southeast regions.

It is noteworthy that Tenório (2003, 2004) has a conciliatory proposal between the hypotheses by Lima (1991, 1999-2000) and Gaspar (1991, 1996). Tenório (2004) argues that an ancient culture of fishermen and gatherer already adapted to coastal zone already existed alongshore maybe since the beginning of Holocene. So she suggests that a variety of archaeological cultures, probably originating from the Continent inland, would be associated with the construction of the sites and that the absence of the typically inland evidence in the sambaquis would be the result of the rapid incorporation of new migrants to a pre-existing culture on the coast.

DISCUSSION AND FINAL REMARKS

Combining the hypotheses suggested for the settlement of fishermen and gatherers in Brazilian sambaquis in the last two centuries, we arrives in, somewhat surprisingly, only two trends, although each has its own implications and internal developments: 1) the populations who built the sambaquis come from the interior of the continent, or 2) they originated from the coast.

The first hypothesis appears as hegemonic in the archeology of sambaquis since the nineteenth century. However, it requires the resolution of a pending issue. Is there a specific spot in the interior of Brazil from where the people who originated the coastal sambaquis would have migrated, or the development of archaeological culture of the sambaquis would have occurred in parallel through different axes inland-coast? The second hypothesis involves the consideration of the origin of the coastal culture. Would it be indigenous or a result of expansion of a coastal culture by the Atlantic route?

Based on the expressive available literature and the synthesis already systematized in Serrano (1946), Emperaire; Laming (1956), Schmitz (1984, 1998), Neves (1988), Prous (1992), DeBlasis et al. (1998), Lima (1999-2000), Tenório (2003), Gaspar et al. (2008) and Wagner et al. (2011), it seems indisputable that there are regional specificities in both the material culture associated to sambaquis and in the biology of human populations buried in them.

Schmitz (1984) situates the first region of dispersion of this archaeological culture between the northern Paraná and São Paulo's south which would have originated from the hunter-gatherers inland (Humaitá Tradition). Neves (1988) points out the valleys of the Rio Ribeira, São Paulo, and Itajaí, Santa Catarina, as probable original access routes to the coast. Lima (1999-2000) reiterates

the proposal and adds Jacuí valley of Rio Grande do Sul. Tenório (2004) suggests three ways: 1) the north of Rio de Janeiro, 2) São Paulo, and 3) the south, coming from Uruguay. Simões (1981) incorporates the ceramic sambaquis of the North and Northeast (Tradition Mina) to the Caribbean ceramic complex. In Rio Grande do Sul, Miller (1969, p. 102), stated that the lithic industries of the sambaquis "fit almost entirely in the collection of Cerrito Dalpiaz, however, the collection of the latter, much more complex, fits only partially in the first (...)".

It seems likely, at the current stage of research, to assume the cultural diversity associated with the construction of coastal sambaquis. The most likely is that the origins have occurred in parallel through various axes inland-coast. It is noteworthy that the practice of intensive exploitation and accumulation of mollusks began in the highlands of the countryside,[11] as shown by the occupation of Capelinha 1 dated of 9250 ± 50 BP. (Figuti et al., 2004).

After settling on the south and southeastern coast of Brazil, the sambaquis builders probably established a series of interactions with one another, along the coast, giving relative uniformity to the material culture identified in the sites. Furthermore, they continued to receive cultural influences and genetic material from both hunter-gatherer populations and the ceramists populations of the countryside. It is perfectly possible to have even been influenced by different groups in the Prata region and in the Amazonian coastline. It should be noted that the shell mounds of North and Northeast regions of the Country correspond too ther archaeological culture and should be considered apart from the settlement of the sambaquis of southern shores of Brazil.

Despite several attempts to resolve the issues highlighted above, there is not, at the current stage of research, a consensus to define a more probable link for the archaeological culture of sambaquis. There seems to be in some respects, overlapping marks of occupations in the "fringe" of coastal settlements, such as the Itaipu Tradition, in Rio de Janeiro, with inland origins, the central portion of Santa Catarina (cf. already mentioned in Neves, 1988) and part of the coast of Rio Grande do Sul. However, the elucidation of these questions depends on an intense research effort and many excavations need to be made so that we can understand its implications in the origin of sambaquis in southern Brazil.

References

BANDEIRA, A. (2006) – O povoamento da América visto a partir dos sambaquis do litoral equatorial amazônico do Brasil. Fundhamentos. São Raimundo Nonato. 7, p. 431-468.

BIGARELLA, J. (1950-1951) – Contribuição ao Estudo dos Sambaquis no Estado do Paraná I, Regiões

---

[11] This hypothesis was already supported by Hurt (1983-1984) and Lynch (1998).

Adjacentes às Baías de Paranaguá e Antonina. Arquivos de Biologia e Tecnologia. Curitiba. 5-6, p. 231-292.

CALDERÓN, V. (1964) – O Sambaqui da Pedra Oca. Salvador: Universidade da Bahia. 88p.

CALIXTO, B. (1904) – Algumas notas e informações sobre a situação dos sambaquis de Itanhaém e de Santos. Revista do Museu Paulista. São Paulo. 6, p. 490-518.

CARDIM, F. (1939[1584]) – Tratados da terra e gente do Brasil. Rio de Janeiro: Brasiliana. 379p.

COSTA. A. (1934) – Introdução à Arqueologia Brasileira: Etnografia e Historia. São Paulo: Nacional. 348p.

DEBLASIS, P. [et al.] (1998) – Some References for the Discussion of Complexity Among the Sambaqui Mound builders from the Southern Shores of Brazil. Revista de Arqueologia Americana. México, D.F. 15, p. 75-106.

EMPERAIRE, J.; LAMING, A. (1956) – Les sambaquis de la côteméridionale du Brésil (campagnes de fouilles 1954-1956). Journal de la Société des Américanistes nouvelle série. Paris. 45, p. 5-123.

EMAPERAIRE, J.; LAMING, A. (1958) – Sambaquis Brésiliens et Amas de Coquilles Fuégiens. México. 2, p. 165-178. Anais do XXXI Congresso Internacional de Americanistas.

FIGUTI, L. [et al.] – [relatório de pesquisa] – (2004) – Investigações arqueológicas e geofísicas dos sambaquis fluviais do vale do rio Ribeira de Iguape. Estado de São Paulo/ FAPESP-1999/12684-2. São Paulo. 106p.

FILIPPINI, J.; EGGERS, S. (2005-2006) – Distância Biológica entre Sambaquieiros Fluviais (Moraes-Vale do Ribeiro-SP) e Construtores de Sítios Litorâneos (Piaçaguera e Tenório-SP e Jabuticabeira II-SC). Revista do MAE. São Paulo. 15-16, p. 165-180.

GASPAR, M. [Tese de doutorado] (1991) – Aspectos da Organização Social de um Grupo de Pescadores, Coletores e Caçadores: Região Compreendida entre a Ilha Grande e o Delta do Paraíba do Sul, Estado do Rio de Janeiro. USP. São Paulo.

GASPAR, M. (1996) – Análise das Datações Radiocarbônicas dos sítios de Pescadores, Coletores e Caçadores. Boletim do MPEG. Belém. 8, p. 81-91.

GASPAR, M. [et al.] (2008) – Sambaqui (Shell Mound) Societies of Coastal Brazil.In Silverman, H.; Isbell, W. eds. – Handbook of South American Archaeology.New York: Springer, p. 319-338.

HARTT, C. (1885) – Contribuições para a Ethnologia do Valle do Amazonas. Archivos do Museu Nacional. Rio de Janeiro. 6, p. 1-174.

HILBERT, P. (1959) – Achados Arqueológicos num Sambaqui do Baixo Amazonas. Instituto de Antropologia e Etnologia do Pará. Belém. 10, p. 1-22.

HURT, W; BLASI, O. (1960) – O sambaqui do Macedo A.52 B., Paraná, Brasil. Conselho de Pesquisa da Universidade do Paraná. Curitiba. 2, 1-98p.

HURT, W. (1983-1984) – Adaptações Marítimas no Brasil. Arquivo do Museu de História Natural. Belo Horizonte. 7-9, p. 61-72.

IHERING, H. (1895) – A Civilização Prehistorica do Brazil Meridional. Revista do Museu Paulista. São Paulo. 1, p. 33-159.

IHERING, H. (1904) – Arqueologia comparativa do Brasil. Revista do Museu Paulista. São Paulo. 6, p. 519-583.

KOSERITZ, C. (1884) – Bosquejos Ethnológicos. Typographia Gundlach & Companhia: Porto Alegre. 83p.

KRONE, R. (1914) – Informações Ethnographicas do Valle do Rio Ribeira de Iguape. In Botelho, C. eds. – Comissão Geográfica e Geológica, Exploração do Rio Ribeira de Iguape. Rothschild & Company: São Paulo, p. 23-34.

LACERDA, J.; PEIXOTO, J. (1976) – Contribuição para o Estudo Anthropologico das Raças Indigenas do Brazil. Archivos do Museu Nacional. Rio de Janeiro. 1, p. 47-75.

LACERDA, J. (1885) – O Homem dos Sambaquis, Contribuição para a Antroologia Brasileira. Archivos do Museu Nacional. Rio de Janeiro. 6, p. 175-203.

LEONARDOS, O. (1938) – Concheiros Naturais e Sambaquis. Ministério da Agricultura. Rio de Janeiro. 109p.

LIMA, T. [Tese de doutorado] (1991) – Dos Mariscos aos Peixes: um Estudo Zooarqueológico de Mudança de Subsistência na Pré-História do Rio de Janeiro. MAE/ USP, São Paulo.

LIMA, T. (1999-2000) – Em Busca dos Frutos do Mar: Os Pescadores-Coletores do Litoral Centro-Sul do Brasil. Revista USP. São Paulo. 44, p. 270-327.

LYNCH, T. (1998) – The Paleoindian and Archaic Stages in South America: Zones of Continuity and Segregation. In Plew, M. ed. – Explorations in American Archaeology: Essays in honor of Wesley Hurt. University Press of America: Lanham, New York, Oxford, p. 89-100.

MADRE DEUS, G. (1920[1797]) – Memórias para a Historia da Capitania de S. Vicente hoje Chamada de São Paulo e Noticias dos Annos em que se Descobrio o Brazil. São Paulo: Weiszflog Irmãos. 240p.

MARTINELLI, S. [Tese de doutorado] (2007) – Processo de Formação do Sambaqui Ilha das Ostras no Litoral Norte do Estado da Bahia. UFBA, Salvador. 130p.

MASI, M. (2001) – Pescadores Coletores da Costa Sul do Brasil. Pesquisas, Antropologia. São Leopoldo. 57, p. 1-136.

MATTOS, A. (1941) A raça de Lagôa Santa: velhos e novos estudos sobre o homem fóssil americano. São Paulo: Nacional. 502p.

MELLO-ALVIM, M. (1978) – Caracterização da Morfologia Craniana das Populações Pré-históricas do Litoral Meridional Brasileiro (Paraná e Santa Catarina). Arquivos de Anatomia e Antropologia. Rio de Janeiro. 3, p. 292-318.

MILLER, E. (1969) – Resultados Preliminares das Escavações no Sítio Pré-Cerâmico RS-LN-1: Cerrito Dalpiaz (abrigo-sob-rocha). Iheringia. Porto Alegre. 1, p. 3-42.

NETTO, L. (1885) – Investigações Sobre a Archeologia Brazileira. Archivos do Museu Nacional. Rio de Janeiro. 6, p. 257-554.

NEVES, W. (1988) – Paleogenética dos grupos pré-históricos do litoral sul do Brasil (Paraná e Santa Catarina). Pesquisas, Antropologia. São Leopoldo. 43, p. 1-176.

NEVES, W. OKUMURA, M. (2005) – Afinidades Biológicas de Grupos Pré-Históricos do Vale do Rio Ribeira de Iguape (SP): uma Análise Preliminar. Revista de Antropologia. São Paulo. 48, p. 525-558.

OKUMURA, M. (2008) – Diversidade Biológica Craniana, Micro-Evolução e Ocupação Pré-Histórica da Costa Brasileira. Pesquisas, Antropologia. São Leopoldo. 66, p. 1-306.

PEROTA, C.; BOTELHO, W. (1994) – Os Sambaquis do Guará e as Variações Climáticas no Holoceno. Revista do Departamento de Geografia, USP. São Paulo. 7, 49-59.

PROUS, A. (1992) – Arqueologia Brasileira. Brasília: UNB. 605p.

RATH, C. (1856) – Fragmentos Geologicos e Geographicos Etc. Typographi Imparcial. São Paulo. 78p.

RAUTH, J. (1962) – O Sambaqui de Saquarema. S.10.B, Paraná, Brasil. Conselho de Pesquisas da Universidade Federal do Paraná. Curitiba. p. 1-75p.

RATHBUN, R. (1878) – Observações Sôbre a Geologia; Aspecto da Ilha de Itaparica, Bahía de Todos os Santos. Archivos do Museu Nacional. Rio de Janeiro. 3, p. 159-183.

RIVERA, M.; ROTHHAMMER, F. (1990) – Relaciones interetnicas entre pueblos de la floresta tropical y areas desertticas del norte de Chile, ca. 5000 A.C.: la tradicion chinchorro. Revista do CEPA. Santa Cruz do Sul. 17, 20, p. 437-456.

ROHR, A. (1962) – Pesquisas paleo-etnográficas na Ilha de Santa Catarina I e sambaquis do litoral sul-catarinense IV. Pesquisas, Antropologia. São Leopoldo. 14, p. 1-48.

ROOSEVELT, A. [et al.] (1991) – Eighth Millennium Pottery from a Prehistoric Shell Midden in the Brazilian Amazon.Science. 254, 13, p. 1621-1624.

SALLES-CUNHA, E. (1963) – História da odontologia no Brasil (1500-1900) Sambaquis – Lagoa Santa – Tupis (aspectos de patologia alvéolo-dentária). Ed. Científica. Rio de Janeiro. 441p.

SAUER, C. (1944) – A Geographic Sketch of Early Man in America. Geographical Review.New York. 34, 4, 529-573.

SCHMITZ, P. (1984) – Caçadores e Coletores da Pré-História do Brasil. UNISINOS, São Leopoldo. 56p.

SCHMITZ, P. (1998) – Peopling of the Seashore of Southern Brazil. In Explorations in American Archaeology: Essays in honor of Wesley Hurt. In Plew, M. ed. – University Press of America: Lanham, New York, Oxford, p. 193-220.

SERRANO, A. (1937) – Subsídios para a Arqueologia do Brasil Meridional. Revista do Arquivo Municipal. São Paulo. 36, 2, p. 5-42.

SERRANO, A. (1946) – The Sambaquis of the Brazilian Coast In Steward, J. ed. – Handbook of South American Indians. United States Government Printing Office: Waschington, p. 401-408.

SILVA, C. [Dissertação de mestrado] (2000) – Herança geológica como ferramenta para a prospecção de sambaquis no litoral norte do Estado da Bahia: o exemplo do sambaqui da ilha das ostras. UFBA, Salvador. 126p.

SIMÕES M. (1981) – Coletores-pescadores ceramistas do litoral do Salgado (Pará). Boletim do MPEG. Belém. 78, 1-26.

SOUZA, A. (1991) – História da Arqueologia Brasileira. Pesquisas, Antropologia. São Leopoldo. 46, 1-157.

TENÓRIO, M. [Tese de doutorado] (2003) – O lugar dos aventureiros: identidade, dinâmica de ocupação, e sistema de trocas no litoral do Rio de Janeiro há 3.500 anos antes do presente. PUCRS, Porto Alegre. 535p.

TENÓRIO, M. (2004) – Identidade Cultural e Origem dos Sambaquis. Revista do MAE. São Paulo. 14, p. 169-178.

TIBURTIUS, G.; BIGARELLA, J. (1960) – Objetos zoomorfos do litoral de Santa Catarina e Paraná. Pesquisas, Antropologia. São Leopoldo. 7, p. 1-51.

UCHÔA, D. (2007[1973]) – Arqueologia de Piaçagüera e Tenório, Análise dos Tipos Líticos de Sítios Pré-Cerâmicos do Litoral Paulista. Hábilis. Erechim. 221p.

WAGNER, G. [Tese de doutorado] (2009a) – Sambaquis da Barreira da Itapeva, uma Perspectiva Geoarqueológica. PUCRS, Porto Alegre. 241p.

WIENER, C. (1876) – Estudos Sobre Sambaquis do Sul do Brazil. Archivos do Museu Nacional. Rio de Janeiro. 1, p. 1-25.

WILLEY, G. (1966) – Introduction to American Archaeology. Vol. I. New Jersey. Prentice-Hall. 530p.

AFFINITY GROUPS FROM THE SHELLMOUND JABUTICABEIRA II (SANTA CATARINA, BRAZIL): WHAT DOES THE CRANIAL MORPHOLOGY SAY?

Mercedes OKUMURA

Museu de Arqueologia e Etnologia, Universidade de São Paulo, Av. Prof. Almeida Prado, 1466, São Paulo, 05508-070
mercedes.okumura@gmail.com

Sabine EGGERS

Laboratório Antropologia Biológica, Depto. de Genética e Biologia Evolutiva, Instituto de Biociências,
Universidade de São Paulo. CP 11461, 05422-970, São Paulo, SP
saeggers@usp.br

Abstract: *The shellmound Jabuticabeira II (Santa Catarina, Brazil) has been considered as a huge late Holocene cemetery, dated from 2880 to 1400 BP. Gaspar et al. (2008) have proposed the existence of an affinity group buried in Locus 2, meaning a group of individuals that was buried in a circumscribed area within a relatively short period of time. Our aim was to test the hypothesis that individuals from Locus 2 presented greater biological affinity in relation to other individuals from the same site. Our results, based on craniometric data from 26 individuals, do not support the abovementioned hypothesis.*

Keywords: *Shellmound, craniometrics, Jabuticabeira II, Brazilian prehistory*

Résumé: *L'amas de coquillage de Jabuticabeira II (Santa Catarina, Brésil) a été considéré comme un vaste cimetière du Holocène tardif, daté entre 2880 et 1400 ans BP. Gaspar et al. (2008) ont proposé l'existence d'un groupe d'affinité enterré dans le Locus 2, soit un groupe d'individus qui a été enterré dans une zone circonscrite dans un laps de temps relativement court. Notre objectif était de tester l'hypothèse que les individus du Locus 2 présentaient une plus grande affinité biologique par rapport à d'autres personnes provenant du même site. Nos résultats, fondés sur des données craniométriques provenant de 26 individus, ne supportent pas cette hypothèse.*

Mots-clés: *Amas de coquillage, craniométrie, Jabuticabeira II, préhistoire brésilienne*

INTRODUCTION

Shellmounds are the most conspicuous archaeological sites in the Brazilian coastal landscape, being known since the sixteenth century by European travelers. Although many naturalists have called attention to the presence of such sites, academic studies aiming to understand the formation and other aspects of these shellmounds only began to be developed in the nineteenth century.

One of the striking features of most shellmounds is the presence of abundant human burials. The large number of skeletal remains exhumed from these sites allowed answering several questions about the biology and the way of life, as well as the origins and dispersions of these coastal groups (Gaspar, 1998, 2000).

The excavation of the shellmound Jabuticabeira II, dated between 2880 and 1400 BP (De Blasis *et al.*, 1998, Assumption, 2010, Giannini *et al.*, 2010) provided important information and allowed for several research projects focusing on very different areas ranging from anthracology (Bianchini *et al.*, 2007; Bianchini, 2008), through geomorphology (Giannini *et al.*, 2010), stratigraphy (Villagran *et al.*, 2010) and to bioarchaeology.

The skeletal remains are abundant in Jabuticabeira II and there is strong evidence that the use of the site as a cemetery could have been related to its construction process. That would happen through the preparation of the burial location and the building of a mound of sediment mixed with shells on top of the corpses (De Blasis *et al.*, 1998, Fish *et al.*, 2000, Gaspar *et al.*, 1999, 2008; Klokler, 2008). Bioarchaeological studies involving the analysis of skeletal material exhumed from the site has revealed new and exciting information on the origins, the biological affinities (Filippini, 2004; Bartolomucci, 2006; Filippini & Eggers, 2005-2006; Okumura, 2008), the health and disease profiles (Storto *et al.*, 1999; Eggers & Okumura, 2005; Petronilho, 2005, Okumura *et al.*, 2007 b) and the subsistence patterns of this group (Boyadjian, 2007; Boyadjian *et al.*, 2007; Okumura & Eggers, 2012).

One of the best characterized burial locations in Jabuticabeira II was called Locus 2 (Figure 1). Gaspar *et al.* (2008:325) describe this location: "A horizontal excavation of 36 m² within a funerary area confirmed that the corresponding dark layers are successive occupation surfaces, sometimes with localized shell pavements and always with numerous postholes in the vicinity of the burials. Posts encircled some burial pits and similarly demarcated whole funerary areas. Additional posts may

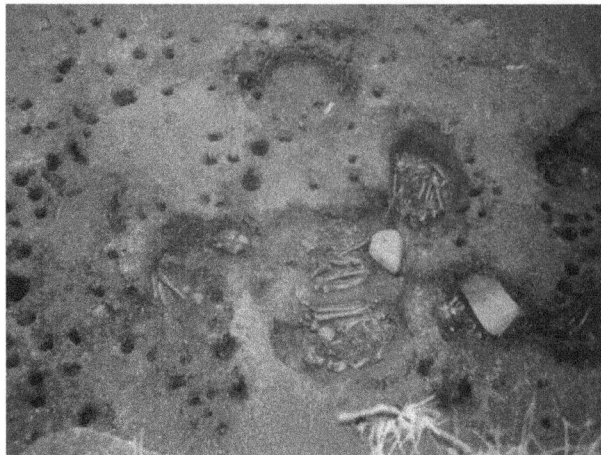

Figure 1. Partial view of Locus 2 presenting five individuals, shell pavements and numerous postholes

have supported miniature structures over graves (as in ethnographic practices), suspended offerings, marked the graves, or served still other purposes." Thus, the individuals buried at Locus 2 have been considered by some researchers as an affinity group; meaning a group of individuals that was buried in an area circumscribed within a relatively short period of time (Gaspar *et al.*, 2008). An individual from the bottom layer and another individual from the top (the funeral layer measures about 25 cm thick) were dated to 2340 ± 50 and 2320 ± 50 yBP, respectively (Beta 188381 and Beta 188382 – De Blasis *et al.*, 2004). The calibrated dates are 2465-2315 for the individual from the bottom and 2370-2180 for the individual from the top (Bianchini, 2008:77). According to Gaspar *et al.* (2008:324-325), areas as Locus 2 could be interpreted as places designated for the burial of specific affinity groups, whose membership would be based on kinship, territorial affiliation or other principles. Such hypothesis raised questions about how this affinity could be investigated. One way would be to check the existence of a greater kinship (translated into cranial morphological similarities) between individuals of Locus 2 in relation to others. This is the aim of our work.

MATERIALS AND METHODS

The sex of skeletal remains was estimated according to the criteria presented by Buikstra & Ubelaker (1994) and White (1991). Only adult individuals were included in the study. The assignment of the age-of-death in adults was performed by verifying the presence of the third molars and the fusion of the spheno-occipital suture.

Craniometric variables are continuous variables obtained from linear measurements (lengths, widths, projections) that are used to characterize the size and shape of the skull (Larsen, 1997). Numerous attempts have been made to separate size from shape, and there is not a single definition of what is exactly size and shape (Lele, 1991; Rao, 2000; Richsmeier *et al.*, 2002). Despite the language used in several studies (including this one), the influence of size can never be completely removed, and size and

shape are not biologically independent, being closely related. However, studies based on skull measurements can benefit from an attempt to partially remove the size factor, especially when the compared groups present important differences in relation to this element. In such cases, if the size factor is not removed, the differences among groups will be produced mostly by size. The use of craniodental elements is quite effective to test hypotheses about phylogenies and taxonomies. This approach reflects the frequent preservation and easy identification of cranial and dental remains, as well as the assumption that the cranial morphology is a good indicator of biological relationships between ancestors and descendents.

Variables and specimens whose missing values exceeded 50% were excluded. This procedure reduced the number of craniometric variables to only six measurements; however, these variables can satisfactorily describe the main dimensions of the skull. The remaining missing values were replaced by averages calculated using all groups. The six craniometric variables have been described by Howells (1973, 1989). They are: GOL (Glabello-occipital lenght), STB (Bistephanic breadth), FMB (bifrontal breadth), DKB (Interorbital breadth), WMH (Minimum malar height), and FMR (Frontomalare anterior radius).

Table 1 presents the sample. We have studied 26 individuals, seven from the Locus 2 and 19 from other loci of the site, which were renamed Locus 1 (see column "Final Locus"). Due to the small sample size, male and female individuals were analyzed together.

Geometric mean was applied to partially remove the size factor (Darroch & Mosimann, 1985). This method divides the value of each variable by the geometric mean of all the variables of the individual or sample. This approach is necessary so that the shape of the skull, and not its size, is responsible for the morphological differences found among groups.

We performed three statistical analyses: Principal Components Analysis, Discriminant Function Analysis and T-Test. The Principal Component Analysis (using covariance matrices) aims to explore the correlations between a large number of interrelated quantitative variables by grouping these variables on a few independent factors or components. The Discriminant Function Analysis locates individuals in groups using a given set of variables. Each individual measurement is compared with the centroid value of each group of individuals, taking into account both the group variance and the number of individuals in each group. The variance is used to determine, within each group, the threshold dispersion of the observed values of each variable. Thus, each individual has a different probability of classification in each group. Individuals present a higher probability of being classified in the largest groups by chance, therefore the probabilities are weighted according to the number of individuals of each group (Albrecht, 1992, Van Vark & Schasma, 1992). The T-

Table 1. Description of the sample, locus, final locus and estimated sex

Identification	Locus	Final locus	Sex
Bur11/E1/L1.25	1.25	1	male
Bur 3B2/E1/L1.05	1.05	1	male
Bur6/E1/L1.10	1.1	1	male
Bur10B/E1/L1.25	1.25	1	male
Bur12C/E1/L1.25	1.25	1	male
Bur15/E1/L1.05	1.05	1	male
Bur17A/E1/L1.05	1.05	1	male
Bur1B/E2/L6C	6C	1	female
Bur2A/E3/L6	6	1	male
Bur42/E3/L1.76	1.76	1	female
Bur43/E3/L1.77	1.77	1	male
Bur107/E5/L1.T18	1/T18	1	male
Bur 3B/L6/QB3/FS2111	6	1	female
Bur 107/L1/T18/E5	1	1	male
Bur 111/112 Pat/Profile E/TL1/L2	TL1/L2	1	male
Bur 111/112 não Pat/Profile E/TL1/L2	TL1/L2	1	male
Bur 115B/L6/FS2136	6	1	male
Bur 118/L6/Layer 2	6	1	male
Bur 121/L6/FS2138	6	1	male
Bur17/E3/L2.05PD	2.05	2	male
Bur25/E3/L2.65	2.65	2	female
Bur41A/E4/L2.05	2.05	2	male
Bur 34/L2.05/E4	2.05	2	female
Bur 36A/L2.05/E4	2.05	2	male
Bur 37/E4/L2.05	2.05	2	male
Bur37/E4	2	2	male

Test verifies if the mean of a variable in a given group is significantly different from the mean observed in another group (Madrigal, 1998:96). Statistical analyzes were performed using the softwares Statistica and SPSS.

RESULTS

Table 2 presents the results generated by the Principal Component Analysis. The first two eigenvalues explain 83.8% of the total variance.

Figure 2 shows the morpho-space formed by the first two Principal Components. It is not possible to verify a clear separation between individuals from Loci 1 and 2. It is also not possible to observe a region on the graph where there is a concentration of individuals of one or another locus.

Table 3 presents the results of Discriminant Function Analysis. Individuals from Locus 1 present a quite high percentage of correct classification, while individuals from Locus 2 do not show the same result, presenting a high percentage of misclassification.

Table 2. Eigenvalues, the total percentage of variance explained by each eigenvalue, the cumulative eigenvalues and the cumulative percentage explained by each eigenvalue

	Eigenvalue	Total % variance	Cumulative Eigenvalue	Cumulative %
1	0.010790	67.49678	0.010790	67.4968
2	0.002606	16.29814	0.013396	83.7949
3	0.001771	11.08097	0.015167	94.8759
4	0.000691	4.32128	0.015858	99.1972
5	0.000127	0.79441	0.015985	99.9916
6	0.000001	0.00841	0.015987	100.0000

Table 4 presents the probability of classification of individuals from Loci 1 and 2. Individuals from Locus 2 are frequently misclassified in Locus 1, presenting relatively high probabilities of classification in that group.

A T-Test was also performed. Table 5 presents the means and standard deviations for each variable at two loci.

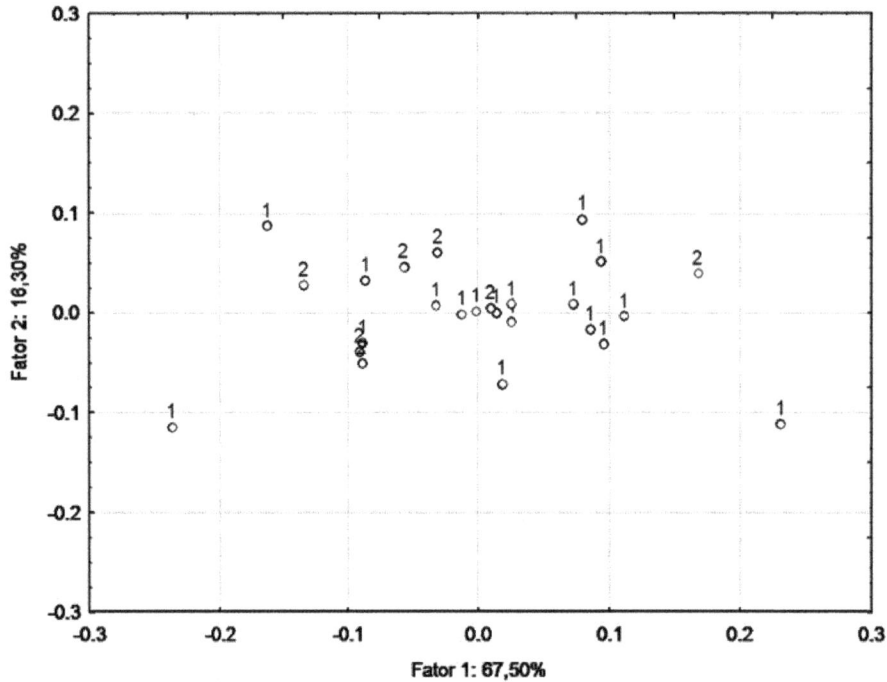

Figure 2. Dispersion of individuals from Loci 1 and 2 in the morpho-space formed by the first two Principal Components

Table 3. Percentage of correct classification and number of individuals classified in Loci 1 and 2

	% of correct classification	Number of individuals classified in Locus 1	Number of individuals classified in Locus 2
Locus 1	94.74	18	1
Locus 2	28.57	5	2
Total	76.91	23	3

Table 6 presents Levene's test, which must be perfomed before the T-Test to check whether or not the variances of the two groups are equal. Since the p values are not significant (column "Levene Sig") equal variances for both groups may be assumed. P values of the T-Test (column "T-Test Sig") presented no significant differences between the group means. The results indicate that the means and standard deviations for each variable in Locus 2 are not significantly different from those observed in Locus 1.

DISCUSSION AND CONCLUSIONS

Our results indicate no significant differences between the proposed affinity group (individuals from Locus 2) and individuals from other loci in Jabuticabeira II. However, this study presents some weaknesses that should be discussed. A major problem was the small sample size, which demanded the analysis of individuals from both sexes together. However, analyzes restricted to the male sample presented very similar results (results not shown in this article). The replacement of the missing values by the mean of the total sample also helped to

homogenize the values. However, this procedure had to be done to obtain a minimum acceptable number of variables and individuals. Furthermore, the unequal sample sizes analyzed by the Discriminant Function Analysis, could have masked possible morphological differences. It is well known that the group with the largest number of individuals is the one that "classifies" most individuals. However, the misclassified individuals from Locus 2 presented a relatively high probability of classification, strongly indicating no important differences between the two groups. Another limitation of this study is the small number of measurements that could be taken from all the skulls to safeguard comparability. This is a consequence of the absence of different cranial fragments, due to problems of preservation and conservation of osteological material. However, the variables included in the analysis represent the major cranial dimensions, being a satisfactory way of describing and comparing these skulls.

Based on our results, it is not possible to say that the individuals buried in Locus 2 present a greater affinity in terms of cranial morphology in comparison to individuals buried in other loci. Although this analysis presents some

Table 4. Probability of classification of individuals from Loci 1 and 2. Asterisk indicates misclassification

Locus	Probability of classification in Locus 1	Probability of classification in Locus 2
1	0.69	0.31
1	0.61	0.39
1	0.68	0.32
1	0.88	0.12
*1	0.35	0.65
1	0.82	0.18
1	0.91	0.09
1	0.74	0.26
1	0.69	0.31
1	0.53	0.47
1	0.81	0.19
1	0.94	0.06
1	0.68	0.32
1	0.96	0.04
1	0.84	0.16
1	0.75	0.25
1	0.96	0.04
1	0.92	0.08
1	0.95	0.05
*2	0.74	0.26
2	0.42	0.58
2	0.32	0.68
*2	0.67	0.33
*2	0.77	0.23
*2	0.68	0.32
*2	0.67	0.33

Table 5. Mean and standard deviation of the six variables for the two loci

Measurement	Locus	Mean	Standard deviation
GOL	1	2.5392	0.0884
	2	2.506	0.0845
STB	1	1.5777	0.0508
	2	1.5704	0.0511
FMB	1	1.4069	0.0406
	2	1.3915	0.0403
DKB	1	0.3186	0.0238
	2	0.3274	0.0196
WMH	1	0.3505	0.0134
	2	0.3578	0.0104
FRC	1	1.5982	0.0652
	2	1.5643	0.0374

limitations, our results corroborate other studies that also found no evidence of affinity groups in the shellmound Jabuticabeira II. The study by Bartolomucci (2006) addressed the same problem using non-metric dental characters. She could not find significant differences between individuals from Locus 2 and the other loci. The same results were obtained by Filippini (2004) through the analysis of non-metric cranial data. In terms of patterns and frequencies of osteo-arthrosis, no major differences regarding intensity and / or activity patterns could differentiate individuals from Locus 2 from other loci (Petronilho, 2005).

Studies focusing on the diet of affinity groups aimed to test the hypothesis of individuals from Locus 2 presenting a different diet, reflecting a differential access to resources. Eight individuals from Jabuticabeira II were analyzed for stable isotopes of Carbon and Nitrogen

Table 6. Levene's test for equality of variances and T-Test

Variable	Levene F	Levene Sig	T-Test	Degrees of freedom	T-Test Sig
GOL	0.004	0.951	0.858	24	0.399
STB	0.132	0.719	0.323	24	0.749
FMB	0.085	0.773	0.861	24	0.398
DKB	0.389	0.539	-0.88	24	0.388
WMH	0.681	0.417	-1.296	24	0.207
FRC	1.238	0.277	1.288	24	0.21

(Richards et al., 2007). The four individuals from Locus 2 (two males and two females) presented less variation between the two markers and a higher average of d15N, suggesting a greater intake of protein in relation to other individuals from Jabuticabeira II. However, Klokler (2008), when including eight further individuals in the sample, found no significant differences. Zooarchaeological analysis on the food offered during funerary rituals for 12 alleged affinity groups from

Jabuticabeira II also could not find any important differences (Klokler, 2008). Likewise, the evaluation of botanic micro-remains extracted from dental calculus has shown no differences in terms of vegetable consumption by individuals from Locus 2 and the other individuals (Boyadjian, 2012). Therefore, none of the approaches presented here (craniometric studies, non-metric cranial and dental analysis, zooarchaeological studies, stable isotopic analysis, patterns and frequencies of osteoarthritis, dental calculus from micro-remains) could reveal any evidence of important differences between the affinity group from Locus 2 and the other loci in Jabuticabeira II.

Finally, it is important to stress the importance of having researchers from many sub-areas of archaeology proposing hypotheses that can be tested through the study of human skeletal remains. Certainly, the analysis of human remains can be greatly benefited from the exploration of archaeological problems proposed by academics whose focus of research is related to issues other than ours.

Acknowledgements

Especial thanks to Maria Dulce Gaspar for proposing the hypothesis of affinity groups. We also would like to thank the coordinators of the excavation of Jabuticabeira II: Maria Dulce Gaspar, Paulo De Blasis, Paul Fish and Suzanne Fish, as well as Cecilia Petronilho and Fabio Parenti for technical help. This research was funded by FAPESP (process 02/13441-0) and CNPq (productivity scholarship for SE). A version of this article in Portuguese has been submitted for publication in Revista do Museu de Arqueologia e Etnologia (MAE-USP).

References

ALBRECHT, G. H. (1992) – Assessing the affinities of fossils using canonical variates and generalized distances. Human Evolution. High Point. 7: 4, p. 49-69.

ASSUNÇÃO, D. C. (2010) – Sambaquis da paleolaguna de Santa Marta: em busca do contexto regional no litoral sul de Santa Catarina. São Paulo: Museu de Arqueologia e Etnologia – USP. 146 p. MSc Dissertation.

BARTOLOMUCCI, L. B. G. (2006) – Variabilidade Biológica entre Sambaquieiros: um estudo de morfologia dentária. São Paulo: Departamento de Genética e Biologia Evolutiva – USP. 105 p. MSc Dissertation.

BIANCHINI, G. F. [et al.] (2007) – Estaca de Lauraceae em contexto funerário (sítio Jaboticabeira-II, Santa Catarina, Brasil). Revista do Museu de Arqueologia e Etnologia. São Paulo. 17, p. 223-229.

BIANCHINI, G. F. (2008) – Fogo e Paisagem: evidências de práticas rituais e construção do ambiente a partir da análise antracológica de um sambaqui no litoral sul de Santa Catarina. Rio de Janeiro: Universidade Federal do Rio de Janeiro (UFRJ). 200 p. MSc Dissertation.

BOYADJIAN, C. H. C. (2007) – Microfósseis contidos no cálculo dentário como evidência do uso de recursos vegetais os sambaquis de Jabuticabeira II (SC) e Moraes (SP). São Paulo: Departamento de Genética e Biologia Evolutiva – USP. 147 p. MSc Dissertation.

BOYADJIAN, C. H. C. [et al.] – Dental wash: a problematic method for extracting microfossils from teeth. Journal of Archaeological Science. 34, p. 1622-1628.

BOYADJIAN, C. H. C. (2012) – Análise e identificação de microvestígios vegetais de cálculo dentário para a reconstrução de dieta sambaquieira: estudo de caso de Jabuticabeira II, SC. São Paulo: Departamento de Genética e Biologia Evolutiva – USP. 226 p. PhD thesis.

BUIKSTRA, J. E.; UBELAKER, D. H. (1994) – Standards for Data Collection from Human Skeletal Remains. Fayetteville: Arkansas Archaeological Survey Research Series nº 44. 272 p.

DARROCH, J. N.; MOSIMANN, J. E. (1985) – Canonical and principal components of shape. Biometrika. Oxford. 72, p. 241-252.

DE BLASIS P. A. D. [et al.] (1998) – Some references for the discussion of complexity among the sambaqui mound builders from the southern shores of Brazil. Revista de Arqueología Americana. Mexico. 15, p. 75-105.

DE BLASIS, P. A. D. [et al.] (2004) – Processos Formativos nos Sambaquis do Camacho, SC: Padrões Funerários e Atividades Cotidianas. Final report. São Paulo: Fapesp (processo: 03/02059-0).

FILIPPINI, J. (2004) – Biodistância entre sambaquieiros fluviais e costeiros. Uma abordagem não-métrica craniana entre três sítios fluviais do Vale do Ribeira – SP (Moraes, Capelinha e Pavão XVI) e três costeiros do sul e sudeste do Brasil (Piaçagüera, Jabuticabeira II e Tenório). São Paulo: Museu de Arqueologia e Etnologia – USP. 247 p. MSc Dissertation.

FILIPPINI, J.; EGGERS, S. (2005/2006) – Distância Biológica entre Sambaquieiros Fluviais (Moraes – Vale do Ribeira – SP) e Construtores de Sítios Litorâneos (Piaçaguera e Tenório-SP e Jabuticabeira II-SC). Revista do Museu de Arqueologia e Etnologia. São Paulo. 15-16, p. 165-180.

FISH, S. K.; [et al.] (2000) – Eventos Incrementais na Construção de Sambaquis, Litoral Sul do Estado de Santa Catarina. Revista do Museu de Arqueologia e Etnologia. São Paulo. 10, p. 69-87.

GASPAR, M. D. (1998) – Considerations of the sambaquis of the Brazilian coast. Antiquity. 72, p. 592-615.

GASPAR, M. D. [et al.] (1999) – Uma Breve História do Projeto de Pesquisa "Padrão de Assentamento e Formação de Sambaquis: Arqueologia e Preservação em Santa Catarina". Revista do Centro de Estudos e Pesquisas de Arqueologia (CEPA). Curitiba. 23: 29, p. 103-117.

GASPAR, M. D. [et al.] (2008) Sambaqui (Shell Mound) Societies of Coastal Brazil. In: Handbook of South American Archaeology (Silverman H, and Isbell WH, eds). New York: Springer. p. 319-335.

GIANINI, P. C. F. [et al.] (2010) – Interações entre evolução sedimentar e ocupação humana pré-histórica na costa centro-sul de Santa Catarina, Brasil. Boletim do Museu Paraense Emílio Goeldi Ciências Humanas. Belém. 15: 1, p. 105-128.

HOWELLS, W. W. (1973) – Cranial Variation in Man: A Study by Multivariate Analysis of Patterns of Difference Among Recent Human Populations. Papers of the Peabody Museum of Archaeology and Ethnology, v. 67. Cambridge: Harvard University Press. 259 p.

HOWELLS, W. W. (1989) – Skull Shapes and the Map: Craniometric Analyses in the Dispersion of Modern Homo. Papers of the Peabody Museum of Archaeology and Ethnology, v. 79. Cambridge: Harvard University Press. 187 p.

KLÖKLER, D. M. (2008) – Food for body and soul: Mortuary ritual in shell mounds (Laguna-Brazil). Arizona: University of Arizona. 369 p. PhD thesis.

LARSEN, C. S. (1997) – Bioarchaeology: interpreting behavior from human skeleton. Cambridge: Cambridge University Press. 461 p.

LELE, S. (1991) – Some comments on coordinate-free and scale-invariant methods in morphometrics. American Journal of Physical Anthropology. 85: 4, p. 407-417.

MADRIGAL, L. (1998) – Statistics for Anthropology. Cambridge: Cambridge University Press. 238 p.

OKUMURA, M. M. M. (2008) – Diversidade morfológica craniana, microevolução e ocupação pré-histórica da costa brasileira. Pesquisas (série Antropologia), vol. 66. São Leopoldo: Instituto Anchietano de Pesquisas. 306 p.

OKUMURA, M. M.; EGGERS, S. (2005) – The people of Jabuticabeira II: reconstruction of the way of life in a Brazilian shellmound. Homo. 55: 3, p. 263-281.

OKUMURA, M. M. [et al.] (2007) – Auditory exostoses as an aquatic activity marker: a comparison of coastal and inland skeletal remains from tropical and subtropical regions of Brazil. American Journal of Physical Anthropology. 132: 4, p. 558-567.

OKUMURA, M. M. [et al.] (2007) – An evaluation of auditory exostoses in 621 prehistoric human skulls from coastal Brazil. Ear, Nose, & Throat Journal. Cleveland. 86: 8, p. 468-472.

OKUMURA, M. M.; EGGERS, S. (2008) – Natural and Cultural Formation Processes on the Archaeological Record: A Case Study regarding Skeletal Remains from a Brazilian Shellmound. In: Archeology Research Trends (Suárez AR, and Vásquez MN, eds). Hauppauge: Nova Science Publishers. p. 1-39.

OKUMURA, M. M.; EGGERS, S. (2012) – Living and eating in coastal southern Brazil during Prehistory: a review. In: Food & Drink in Archaeology 3 (Collard D, Morris J, and Perego E, eds). Totnes: Prospect Books. p. 55-64.

PETRONILHO, C. C. (2005) – Comprometimento articular como um marcador de atividades em um grande Sambaqui-Cemitério. São Paulo: Departamento de Genética e Biologia Evolutiva – USP. 176 p. MSc Dissertation.

RAO, C. (2000) – A note on statistical analysis of shape through triangulation of landmarks. Proceedings of National Academy of Science (PNAS). 97: 7, p. 2995-2998.

RICHARDS, M. [et al.] (2007) – Stable Isotopes and what they reveal about paleodiet in Jabuticabeira II. XIV Congresso da Sociedade de Arqueologia Brasileira, Florianópolis. Simpósio "Bioarqueologia de Sambaquis".

RICHTSMEIER, J. T. [et al.] (2002) – The Promise of Geometric Morphometrics. American Journal of Physical Anthropology. suppl 45: 63-91.

STORTO, C. [et al.] (1999) – Estudo preliminar das paleopatologias da população do sambaqui Jaboticabeira II, Jaguaruna, SC. Revista do Museu de Arqueologia e Etnologia. São Paulo. 9, p. 61-71.

VAN VARK, G. N.; SCHAFSMA, W. (1992) – Advances in quantitative analysis of skeletal morphology. In: Skeletal biology of past people: research methods (Saunders, S. R.; Katzenberg, A. eds). New York: Willey-Liss. p. 225-257.

VILLAGRAN, X. S. [et al.] (2010) – Lecturas Estratigráficas: Arquitetura Funerária y Depositación de Residuos en el Sambaquí Jabuticabeira II. Latin American Antiquity. 21: 2, p. 195-216.

WHITE, T. D. (1991) – Human Osteology. San Diego: Academic Press. 455 p.

SUBSTRATS NEOLITHIQUES AUX ARTS
TRADITIONNELS DES BALKANS

Marcel OTTE

Professeur de Préhistoire, Université de Liège

Résumé: *Les civilisations paysannes des Balkans actuels possèdent une mythologie populaire et un art artisanal extrêmement riches, à forte persistance traditionnelle. L'effet de la christianisation ne se fait sentir que très superficiellement, en particulier par la superposition des fêtes calendaires aux actions de certains saints et aux cultes de la Vierge. La signification profonde de ces rituels, encore bien connue, coïncide d'ailleurs avec les rythmes saisonniers. Les motifs décoratifs actuels les rappellent clairement, en passant des thèmes figurés (arbres, chevaux, serpents, par exemple) aux schémas par dérive plastique. Les motifs religieux originels (« païens ») se maintiennent donc au titre d'éléments « décoratifs », parfois même à l'insu des paysans qui les utilisent encore. On peut donc y « lire » les motifs fondamentaux articulés par la métaphysique néolithique régionale, elle aussi extrêmement puissante et de beaucoup plus longue durée que la christianisation. Les mêmes défis traversent d'ailleurs la paysannerie balkanique qu'au néolithique apparemment forgé sur place.*

Abstract: *The rural civilizations of the modern Balkans possess a popular mythology and an extremely rich craft art with a strong traditional persistence. The effect of Christianization is felt only very superficially, in particular by the superposition of annual festivals with the actions of certain saints and cults of the Virgin Mary. The profound significance of these rituals, still well known, also coincides with seasonal rhythms. Modern decorative motifs clearly recall them, passing from figured themes (trees, horses, snakes, for example) to plastic patterns. Original religious motifs ("pagan") are thus maintained as "decorative" elements, sometimes even without the knowledge of the peasants who still use them. We can thus "read" the fundamental motifs articulated by the regional Neolithic metaphysics, itself extremely powerful and of much longer duration than Christianization. The same challenges moreover traverse the Balkan peasantry that the Neolithic apparently forged in place.*

Comme l'ont brillamment démontré Mircea Éliade (1976) et Claude Lévy-Strauss (1962), les articulations religieuses se placent à la jonction des coutumes traditionnelles et des phénomènes cycliques où ils s'inscrivent. Les coutumes constituent des ensembles de relations symboliques choisies par un peuple dans un certain milieu. Ces systèmes de relations sont si puissants qu'ils se maintiennent quasi-intacts malgré l'écoulement du temps ou les transformations du milieu. Leur ultime justification se place en effet dans la sphère métaphysique et accroche les règles de la vie sociale aux raisons même de l'existence. Ainsi, via ses symboles, la vie sociale porte un sens et peut être poursuivie, garantie, justifiée, pour soi-même comme par rapport aux rythmes naturels, tels les mouvements astraux, le rythme des saisons, les cycles de naissance et de mort. Un « ordre » est dès lors apporté à la fois à l'existence de l'univers et aux fonctions sociales qui s'y accordent. L'exclusion ou la mise hors circuit, plus ou moins explicite, d'individus enfreignant ces règles porte dès lors une sorte de valeur sacrée correspondant inconsciemment au maintien de cet ordre grâce auquel le groupe aurait « toujours » survécu. Le conditionnel s'impose ici car autant les fonctionnements traditionnels peuvent être variés dans le même milieu, autant les « marginaux » ainsi définis peuvent-ils, au risque de leur vie, faire admettre des lois du changement et le déterminisme futile d'une civilisation.

Cet ordre existe de manière si puissante à travers les millénaires qu'on peut en suivre la trace par les expressions matérielles laissées par les rituels qui le perpétuent. Ces expressions apparaissent avec d'autant plus de netteté qu'elles s'appliquent à des phénomènes accessoires, là où l'emprise de la technicité laisse le champ libre aux harmonies plastiques, tels les costumes, les décors muraux, les textiles éphémères ou les masques d'usages occasionnels. Ces reflets lointains d'une métaphysique disparue se perpétuent discrètement grâce à une pirouette propre aux mécanismes historiques: ils sont rangés parmi les activités « folkloriques » c'est-à-dire extraits en quelque sorte de l'histoire en marche: ils sont tous devenus « marginaux ». Par l'ironie du temps, ce qui fut sacré devient dérisoire, ou considéré comme tel.

Les manifestations mythiques les plus fondamentales, celles qui renouent les liens entre la société et l'univers, furent le plus souvent exprimées par une gestuelle et par des discours éphémères: les voies du sacré répugnent à la fixité et à l'accessibilité qu'impose l'écriture. Elles doivent rester secrètes afin de garder toute leur force et leur cohésion. Transmises oralement, elles s'inscrivent dans les coutumes et y prennent racine, puis se développent, à l'abri de toute remise en cause. Mais chaque manifestation ritualisée s'exprime aussi par la vue où viennent jouer les expressions plastiques, autant chargées de sacralité que le rituel lui-même, mais le rendant en quelque sorte plus puissant par sa permanence. C'est ainsi que si, par exemple, les rites agraires possèdent des structures analogues de la Chine à la Méditerranée et jusqu'aux Amériques, l'infinie variation des thèmes décoratifs qui les accompagnent permet à cour sûr de distinguer les régions et les traditions au premier regard. Car, le plus souvent, le poids des valeurs pèse davantage sur les variations stylistiques que sur les structures universelles: c'est en effet là où se réfugie la singularité d'un groupe, donc l'auto-reconnaissance individuelle.

Le cas des Balkans néolithiques est spécialement éloquent car les sociétés y sont puissantes, anciennes, longues et diversifiées (Gimbutas, 1991; Kozlowski, 1993), de telle sorte que leurs valeurs imprègnent encore les campagnes où elles furent parfois approchées par des études ethnographiques récentes. Dès lors, la mise en perspective des symboles graphiques néolithiques y gagne une infinie richesse.

LES RYTHMES SAISONNIERS

L'étude de Nicolaï Nikov (2004) illustre parfaitement cette relation entre les rythmes naturels et les rituels sociaux préservés dans les campagnes de l'actuelle Bulgarie. L'année s'amorce par un jeu de masques aux références animales et agressives, telles les cornes de béliers. Ils incarnent et expriment la rudesse de l'hiver à combattre. Des rameaux d'arbustes décorés y sont opposés, intermédiaires entre sacré et réel, ils rétablissent la fécondité. Durant les débuts de l'année solaire, l'eau est glorifiée pour sa force purificatrice et magique, sous la forme d'une croix jetée à la rivière. C'est aussi le moment de la fraternité au sein du groupe, spécialement les grands-mères accoucheuses. Ensuite, les rituels portent sur le renouveau du soleil levant et de la renaissance des défunts. Les relations parentales sont ensuite renouvelées au printemps. Les animaux de trait (bovidés) sont honorés. Des feux sont allumés et des flèches lancées. Cette période est aussi celle des masques et déguisements pour chasser les mauvais esprits des champs et les rendre fertiles. Les chevaux sont ensuite honorés afin d'assurer leur santé et fertilité. Au mois de mars, les champs sont ensemencés après y avoir enfoncé des branches incandescentes. Le retour du soleil est annoncé par l'arrivée des hirondelles et des cigognes, accompagnées par des champs et des sacrifices. Le mois d'avril est dédié aux femmes non mariées qui exhibent leurs plus beaux apparats, en insistant sur la couronne, le décolleté et le tablier. En avril, les dragons viennent tourmenter les jeunes filles, qui sortent vainqueurs évidemment de ces « combats »... Les œufs sont peints et les deux premiers pondus sont teints en rouge, couleur du dieu des foudres; ils sont gardés toute l'année car chargés de vertus curatives. À nouveau, les décors des œufs peints évoquent les formes dominantes de la décoration: l'arbre floral établissant le lien avec les cieux, les croix, les spirales et les fleurs. Le mois de mai est consacré aux serpents, animaux redoutés, étranges et dissimulés dans les sols. Ils en sont chassés par des martèlements répétés sur les récipients. La grêle est redoutée aussi dans ce mois et écartée par des tirs sur les aigles, censés l'apporter aux champs ensemencés. La grêle est aussi écartée par des statuettes d'argile, des deux sexes, enterrées aux entrées des villages pour en interdire l'accès au dieu de la grêle. Les défunts sont alors commémorés par des ablutions d'eau et de vin et des dépôts de feuilles de noyers. Au mois de juin, une sorte d'initiation sélectionne les jeunes hommes, honnêtes, sains et forts. Ils sont isolés et laissés à jeun durant une semaine. Les danses et les fêtes consacrées à la bonne récolte se succèdent aux sons de la flûte et du tambour. Le calendrier mythologique divise l'année au solstice d'été, cette date est marquée par de nombreux rituels. Le mois de juin est parcouru par de nombreuses festivités afin de protéger et d'augmenter la moisson. En juillet, les malades vont aux eaux curatives et y déposent des bouts de vêtements, censés emporter la maladie. À nouveau, un arbrisseau est dressé afin d'y accrocher ces colifichets, d'où la maladie s'envole, emportée par le vent. Le mois de juillet est consacré au dieu de la foudre et du tonnerre, afin qu'il épargne les récoltes des ravages provoqués par les torrents et les inondations. Au mois d'août, consacré aux récoltes, les fêtes s'y succèdent, telle celle de l'adieu aux champs durant laquelle les derniers épis sont solennellement fauchés par de jeunes filles. Ils sont arrosés d'eau fraîche et fertilisés avec du pain. À la fin du mois d'août, les blés sont engrenés rituellement sur les aires battues par des bœufs. À la même période, le jour est égal à la nuit, l'eau est également coupée et on ne peut plus se baigner: les dragons rentrent chez eux. En septembre, les eaux sont « bénites », mêlées à la farine, elles forment des galettes destinées à nourrir les bœufs qui entament l'ouverture des premiers sillons, où les œufs sacralisés seront déposés. En septembre, des galettes en forme de croix sont façonnées et enterrées devant des poteaux. Les malades sacrifient une poupée enduite de miel et couronnée. En octobre, les journaliers rentrent chez eux, avec un bélier, les maisons en cours de construction sont consacrées par le sacrifice d'un mouton. En novembre, les fêtes sont consacrées aux défunts, les tombes arrosées, un repas y est organisé où chacun laisse une part aux défunts, afin qu'il « vive » dans l'au-delà. Le mois de novembre est consacré à honorer le loup, puis l'ours, où des rituels écartent les mauvais sorts que ces animaux dangereux pourraient faire aux villageois. En décembre débutent les fêtes pour saluer la nouvelle année, par d'autres rituels liés à la purification des étables et toute une série d'actions destinées à faire fructifier les futures récoltes, le bétail et les hommes. C'est aussi le mois des morts et de la renaissance: un porc est sacrifié car ses entrailles sont assimilées à l'utérus de la moisson. De la fin décembre au début janvier, on évite de quitter la maison car le diable rôde. Des êtres maléfiques (vampires) cherchent à sucer le sang des moissons et la fertilité des sols. L'ail, cousu aux vêtements, est utilisé pour ses vertus protectrices.

Chacune de ces cérémonies requiert les costumes et les coiffes appropriées où les signes de leur consécration sont brodés. L'arbre de vie y est spécialement fréquent, ainsi que les spirales, signes de perpétuité. Les masques, les couleurs, les gâteaux, les vaisselles employées, tous portent la marque de leurs fonctions magiques étalées au fil de l'année. Les libations sont fréquentes, et ne laissent aucune trace. Les choix des fleurs, des feuilles et des branches reflètent le sens de ces fêtes. Les broderies de tabliers portent des schémas issus des animaux honorés ou redoutés: serpents ou béliers. Les œufs décorés, les couronnes de fleurs, au symbolisme puissant ne laissent subsister aucune trace également, pas plus que les feux, les gestes et les danses pourtant si fervents et chargés de pouvoirs magiques, de signification protectrice ou plein d'espérance. Ce cas particulier nous ramène aux situations analogues, vécues en préhistoire dont pourtant les fondements économiques et les cultes astraux furent si proches.

RICHESSE ET SYMBOLES DES COSTUMES ACTUELS

Un ouvrage spécial fut consacré aux diversités des décors vestimentaires, cérémoniels et aux caractères fortement traditionnels (Komitska et Borissova, 2000). Les symboles y abondent à foison et on y retrouve facilement la relation aux sens des fêtes: l'oiseau qui emporte les vœux, les cercles solaires, les sinuosités des serpents et, surtout, l'image ambiguë de l'arbre sacré liant la terre et le ciel, mais dont la silhouette évoque tout autant un homme aux bras dressés. En outre, ces costumes sont absolument somptueux. Par leur fraîcheur et portés avec fierté, ils restituent un pan de la conscience religieuse néolithique régionale: le même « esprit » y règne, dans les rapports texturaux, le jeu des formes amples, le choix des couleurs, les jeux des schémas. Ils disent comment la foi s'est transformée en formes. Ils expriment à quel point l'expression d'une appartenance ethnique se matérialise dans l'élégance. Ils montrent l'aspect vital des cérémonies à respecter rigoureusement. Mais ils désignent aussi l'importance de la perte du document archéologique qui a perdu la brillance, les couleurs, et, surtout, la fierté dégagée par les regards des femmes ainsi sacralisées.

La signification religieuse des signes à ce point schématisés était peut-être déjà perdue dès la préhistoire: ils étaient devenus, comme nos lettres de l'alphabet, de pures formes abstraites mais investis de valeur sacrée, codée, significative. Ils donnent une leçon d'humilité, autant sur nos créations artistiques contemporaines que sur la splendeur fanée des documents retrouvés en préhistoire.

ARTISANATS

Avec la même constance (Kovacheva-Kostadinova *et al.*, 1994), on voit réapparaître les motifs mythologiques sur les ustensiles d'usages les plus banals. De la boîte à la cuillère, des masques rituels aux décors des cannes aux crosses de bergers, les thèmes d'animaux mythiques, à peine schématisés, traversent les contraintes mécaniques les plus diverses, autant que les nécessités fonctionnelles, d'apparence si contraignantes. La même symétrie et les mêmes accords de tons structurent les décors brodés des étoffes familières, nappes, couvertures, décorations murales. Une grammaire semble s'imposer aux formes aussitôt que la contrainte technique est satisfaite et laisse libre cours aux « décors secondaires », telles les tasses en bois, ou les têtes de quenouilles, les poires à poudre ou les chandeliers. Toujours, la rosace domine, l'arbre sacré s'impose et les spirales s'enroulent telle la vitalité spontanée de l'univers et des hommes. Les monstres et

les masques surgissent du décor, sinueux et coloré. Jusque dans les décors les plus contraints par les lois mécaniques, les plus dépouillés, tels les croix et les losanges, la même harmonie réunit les teintes, les textures, les rythmes, comme s'ils transperçaient les nécessités fonctionnelles pour rejoindre les mondes obscurs des symboles oubliés dont seuls les schémas démystifiés subsistaient.

QUELQUES EXEMPLES COMPARATIFS AU NEOLITHIQUE

La documentation consacrée aux mêmes effets de style se trouve dans une littérature archéologique mille fois plus importante, paradoxalement, que celle consacrée aux récits mythologiques contemporains. L'abondance des fouilles, la richesse de ces civilisations, leur immense diversité et la durée multi-millénaire de ces productions expliquent et justifient cette abondance de l'information archéologique issue des Balkans néolithiques. Parmi l'ample moisson de travaux édités, nous n'avons sélectionné ici que quelques synthèses significatives (Kruta, 1992; Gimbutas, 1991). Tout comme dans les récits mythiques actuels, certains thèmes fixes abondent, telle la spirale et les méandres. Le plus souvent, les schémas décoratifs sont portés sur des statuettes, essentiellement féminines, comme s'il s'agissait de motifs portés aujourd'hui sur les robes et les costumes rituels. La relation à l'eau y est aussi constante par exemple par les statuettes où s'assemblent le récipient et la figurine, ou les méandres et les spirales. La schématisation y est du même style qu'aujourd'hui, sur la vaisselle ou sur les robes, bien qu'il s'agisse ici de contraintes mécaniques toutes différentes (peintures ou gravures sur statuettes). Les thèmes des oiseaux (ascension des vœux) et des serpents (fertilité et dangerosité des sous-sols) figurent non seulement au titre de décoration mais aussi comme statuettes, ou les deux animaux, aux fonctions opposées mais complémentaires, furent parfois associés. Les béliers, si souvent employés comme thèmes agressifs dans les masques rituels actuels, abondent aussi sous forme de statuettes ou de décors gravés. Les losanges horizontaux figurent autant sur les robes des statuettes d'argile que sur les robes et tabliers actuels. Bien entendu, l'image de la femme est abondamment reproduite, dans toutes les positions, y comprises celles d'accouchement, de « déesses » trônant ou hiératiques, figées comme en réponse à une pose fixée. Elles rappellent alors toutes les images où les femmes actuelles, dès qu'elles portent les costumes rituels, se dressent et exposent les attributs de leur fonction sacrée, comme une affirmation, une exhibition de leur raison d'être et comme porteuses du destin du groupe. Les monstres, démons et êtres composites annoncent l'esprit du « dragon » des contes populaires actuels, courant la nuit dans les champs morts, au cœur de l'hiver. Les taureaux et les oiseaux rapaces, si courants dans les rituels ethnographiques, forment la base de décors et de modelages dès le néolithique. Des statuettes, interprétées comme symboles de mort et de résurrection par l'opposition symétrique de leur décor, évoquent la célébration coutumière de cette phase au creux de l'hiver solaire (solstice). Les œufs, sacralisés aujourd'hui (peinture et offrandes aux sillons), chargés de symbolique spontanée de la reproduction en germe, sont aussi un motif fréquent de décors gravés ou peints.

CONCLUSION

Ainsi, trop souvent réduites au niveau anecdotique, les activités dites « folkloriques » reflètent en fait des formes d'équilibre multi-millénaires qu'une société s'est structurée pour se forger une identité. Les repères fondamentaux de la conscience collective y restent liés à des schémas dont le sens originel fut le plus souvent perdu, en dépit d'une répétition perpétuelle. Les fêtes, les rituels, les costumes s'articulent aux rythmes des phénomènes saisonniers, des moments-clefs des activités économiques, des cycles biologiques de la vie (mariages, naissances, morts et rappels des défunts). Tout cela fonctionne en harmonie avec les mouvements célestes, les jeux des planètes et des étoiles. Autant de solutions forgées aux palpitations de l'existence, terrestres, célestes et humaines, ont déterminé des formules, considérées comme « païennes » mais dont les valeurs n'ont jamais vraiment quitté l'esprit d'un peuple. Si on y est sensible et attentif, ces formules s'imposent à l'évidence, tant leur goût est prononcé et largement distribué. Dans le cas des Balkans néolithiques, la permanence des signes fut si puissante qu'elle absorba les modifications sociales inévitables lors des introductions successives des métaux (or, cuivre, bronze, fer) car elles n'altéraient pas en profondeur les fondements d'une économie restée majoritairement agricole. C'est aussi pour cela que leurs témoignages actuels s'expriment principalement dans les campagnes. Les activités urbaines ont brisé le statut mythique des décors et des masques: elles les ont transformées en une matière morte, abandonnée sous la poussière des musées et de l'oubli. Des pratiques rituelles renouvelées dans les rues de Sofia, de Bucarest ou de Belgrade, ne seraient que des curiosités sans âme. Les mêmes, transposées en contexte campagnard susciteraient une sourde nostalgie et y acquerraient aussitôt le prestige d'une ambiance en résonance avec une foi profonde. Les motifs chrétiens, comme superposés au titre de prétexte, y constituent, à l'inverse, des singularités immédiatement repérées par leur contraste sur un fond dominé par l'esprit traditionnel. Il est d'ailleurs significatif que, parmi tous les thèmes offerts par la chrétienté, celui qui y fut prélevé préférentiellement est celui de saint Georges dominant le dragon, comme s'il entrait plus facilement que d'autres dans une mythologie païenne où le dragon combat les forces de la terre afin de les rendre stériles. Cet être composite, formé par les composantes d'animaux dangereux (serpent, félidé, rapace) rappelle d'ailleurs l'exact équivalent du néolithique chinois où il incarne les forces du mal à combattre annuellement à dates fixes. Lui aussi, d'héritage lointain, semble incarner le vestige du voyage extatique du shaman paléolithique. Il se superpose, dans les deux cas, aux expressions naturelles maîtrisées, tels les taureaux, les béliers et les chevaux de trait. La solide harmonie des millénaires prospères du

néolithique y oppose aussi les thèmes astraux (soleils, étoiles, cycles saisonniers) comme si les rituels d'organisation rythmique en récupéraient la force, la constance et la régularité. En ethnographie, comme en préhistoire balkanique, tout se passe comme si les rituels et les signes exhibés précédaient et annonçaient les évènements astronomiques, plutôt que de s'y confondre. Ce « décalage » chronologique et symbolique, soigneusement calculé et maintenu, implique que ces rituels provoquent les mouvements saisonniers plutôt que les célébrer. Telle est la force réelle des rituels périodiques et la raison pour laquelle les rythmes du mariage ou de la consécration des morts sont si strictement définis. Par cette mise en harmonie céleste, les consécrations humaines se rattachent aux cycles astronomiques perpétuels. Ainsi, la vie sociale, comme la vie économique se trouvent-elles garanties par les forces célestes. C'est pourquoi aussi les symboles de cette harmonie persistent avec une telle puissance jusqu'à nous et pourquoi aussi leur revitalisation actuelle laisse froid le citadin et comble d'émotion le paysan. Les voies de la mythologie ethnographique restituent la vocation religieuse des arts protohistoriques, en leur rendant un peu d'âme et de dignité. Inversement, la profondeur prise sur le temps par les documents archéologiques, offre une garantie de stabilité aux éléments du « folklore » actuel. Globalement enfin, si cette référence à la continuité des symboles, des signes et des rites était remise en cause, l'ethnographie actuelle s'ouvrirait sur une béance absurde, car elle deviendrait sans passé donc sans signification. Si on accepte de considérer la cohérence de l'esprit humain, y compris dans la permanence de sa quête métaphysique, alors les sources matérielles livrées par l'archéologie doivent trouver un écho dans les pratiques actuelles. Parmi les plus profondes convictions se placent les rapports aux cycles perpétuels de l'univers et, plus discrètement, tous les reflets décoratifs de cette harmonie intellectuelle. Directement sensibles à l'âme et en écho à la métaphysique, les harmonies esthétiques en restituent aussitôt la saveur, comme l'âme d'un peuple restituée, portée par les générations successives qui y ont trouvé leurs raisons d'exister.

Bibliographie

ÉLIADE, Mircea (1976) – *Histoire des croyances et des idées religieuses*, Paris, Payot.

GIMBUTAS, Marija (1991) – *The Language of the Goddess*, San Francisco, Harper.

KOMITSKA, Anita; BORISSOVA, Veska (2000) – *Bulgarian folk costumes*, Sofia, Borina Publishing House.

KOVACHEVA-KOSTADINOVA, Vyara; SARAFONA, Maria; CHERKEZOVA, Marina; TENEVA, Nadezka (1994) – *Traditional Bulgarian costumes and folk arts*, Sofia, Bulgarian Academy of Sciences.

KOZLOWSKI (dir.), Janusz (1993) – *Atlas du Néolithique Européen. L'Europe orientale*, vol. 1, ERAUL 45, Liège.

KRUTA, Venceslas (1992) – *L'Europe des origines. La Protohistoire 6000-500 avant J.-C.*, Paris, Gallimard.

LÉVI-STRAUSS, Claude (1962) – *La Pensée sauvage*, Paris, Plon.

NIKOV, Nikolaï (2004) – *Les fêtes bulgares à travers les mythes et les légendes*, Sofia, Musée National du Livre.

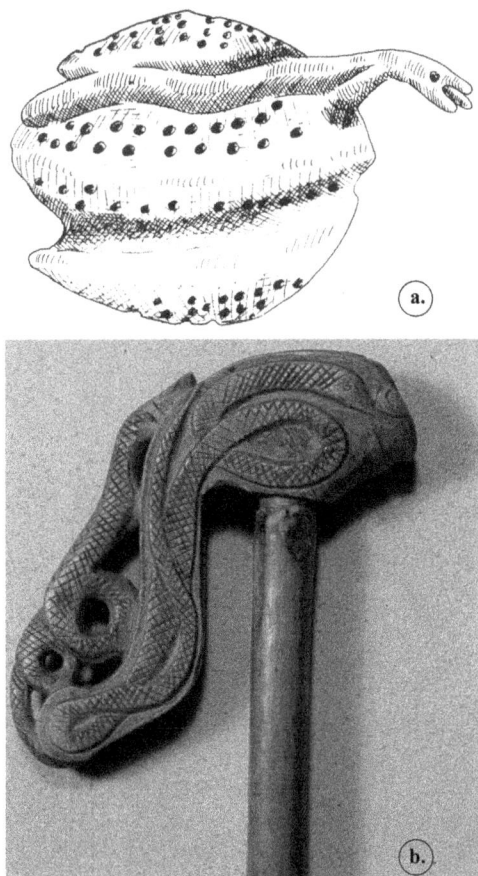

Figure 1. Le thème du serpent apparaît dans le folklore sous la forme de forces terrestres. Ici sous le manche de canne (b) et dès le néolithique sous la forme d'anse de vase (a). (a) Gimbutas, 1991. p. 49, fig. 85; (b) Kovacheva-Kostadinova et al., 1994, p. 78, fig. 138

Figure 2. Le thème de l'arbre de vie perpétuelle apparaît dès le néolithique. Il se retrouve sous forme schématique dans le folklore actuel (a, b et d) et sous forme réaliste lors des festivités (c). (a) Komitska, Borissova, 2000; (b) Nikov, 2004, p. 95; (c) Kovacheva-Kostadinova et al., 1994, p. 84, fig. 159; (d) Kovacheva-Kostadinova et al., 1994, p. 80, fig. 144

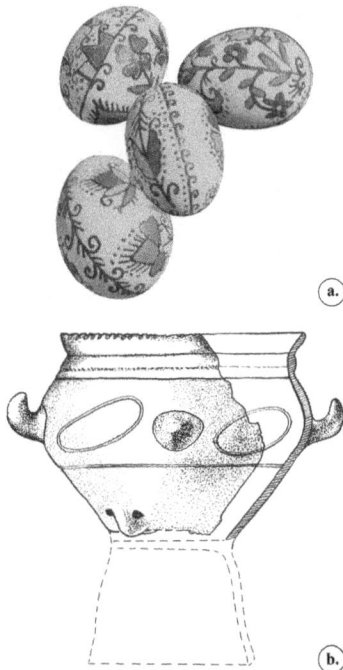

Figure 3. L'œuf, comme symbole de vie ultérieure, est un thème chargé d'espoir de vie dès le néolithique (b) et toujours dans le folklore actuel (a). (a) Nikov, 2004, p. 39; (b) Gimbutas, 1991, p. 215, fig. 333

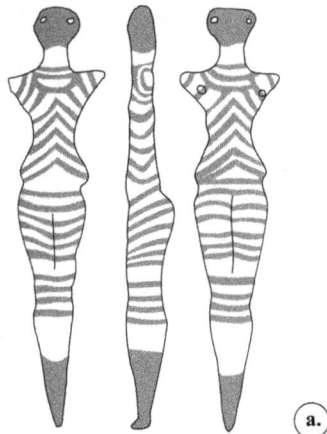

Figure 5. Le thème du cheval correspond dans le folklore actuel à la force de traction de l'araire (c), si fondamentale pour le renouvellement de la vie. On le voit sous forme réaliste ou schématique dans le folklore actuel (a et b). (a) Kovacheva-Kostadinova et al., 1994, p. 45, fig. 32; (b) Kovacheva-Kostadinova et al., 1994, p. 80, fig. 143; (c) Nikov, 2004, p. 29

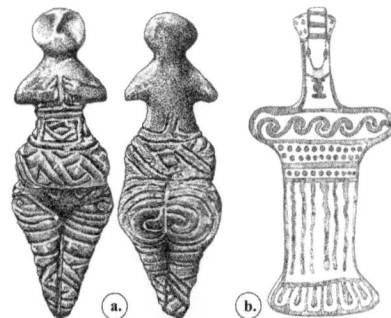

Figure 4. Les motifs décoratifs observés sur les statuettes néolithiques (a) évoquent les costumes traditionnels des campagnes bulgares (b). (a) Gimbutas, 1991, p. 173, fig. 274; (b) Nikov, 2004, p. 37

Figure 6. Le thème de la spirale, signe de vie, de renouvellement par l'eau et la renaissance, se trouve aussi bien au néolithique (a et b) que dans les robes actuelles (c). (a) Gimbutas, 1991, p. 165, fig. 257; (b) Gimbutas, 1991, p. 131, fig. 213; (c) Kovacheva-Kostadinova et al., 1994, p. 82, fig. 152

*Figure 7. Le thème de la femme décorée
apparaît sur les céramiques néolithiques
peintes (a) et sur les robes actuelles (b).
(a) Gimbutas, 1991. p. 239, fig. 373;
(b) Kovacheva-Kostadinova et al.,
1994, p. 50, fig. 49*

*Figure 9. La structure du décor vestimentaire traverse tous les
temps en marquant la distinction entre le plastron et la jupe, et en
insistant sur les motifs rayonnants, spiralés et en étoile. (a)
Gimbutas, 1991, p. 205, fig. 323; (b) Kruta, 1992, p. 144, fig. 109;
(c) Kruta, 1992, p. 143, fig. 108; (d) Komitska, Borissova, 2000*

*Figure 8. Le thème de l'orant aux bras
dressés se retrouve comme un fidèle qui
implore ses dieux. (a) Kovacheva-
Kostadinova et al., 1994, p. 66, fig. 96;
(b) Kovacheva-Kostadinova et al., 1994,
p. 70, fig. 107; (c) Nikov, 2004, p. 12*

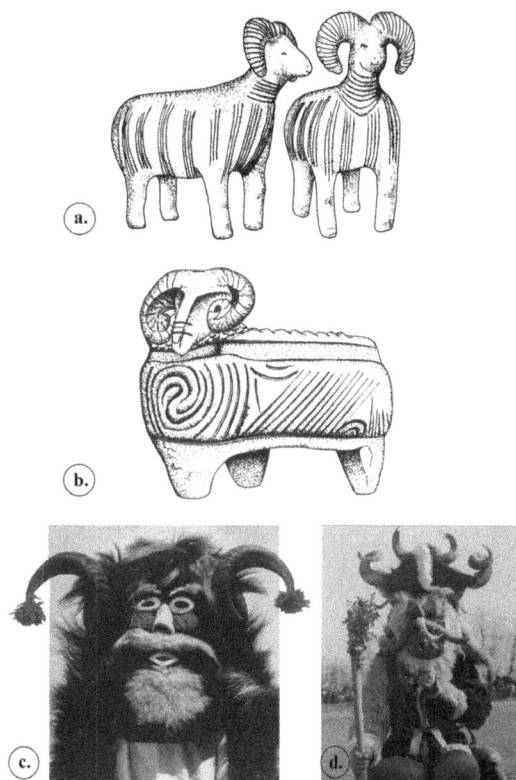

*Figure 10. Le thème du bélier, symbole de la force animale vaincue,
apparaît dès le néolithique (a et b) et se poursuit dans les fêtes du
printemps actuelles sous forme de masques (c et d). (a) Gimbutas,
1991, p. 78, fig. 123; (b) Gimbutas, 1991, p. 77, fig. 121; (c) Nikov,
2004, p. 26; (d) Kovacheva-Kostadinova et al., 1994, p. 85, fig. 161*

Figure 12. La vaztika est un signe de vitalité par l'empennage donné à la croix qui lui donne son mouvement. Il se retrouve à l'identique sur les décors des robes néolithiques (a) et des robes actuelles (b). (a) Kruta, 1992, p. 142, fig. 107; (b) Komitska, Borissova, 2000

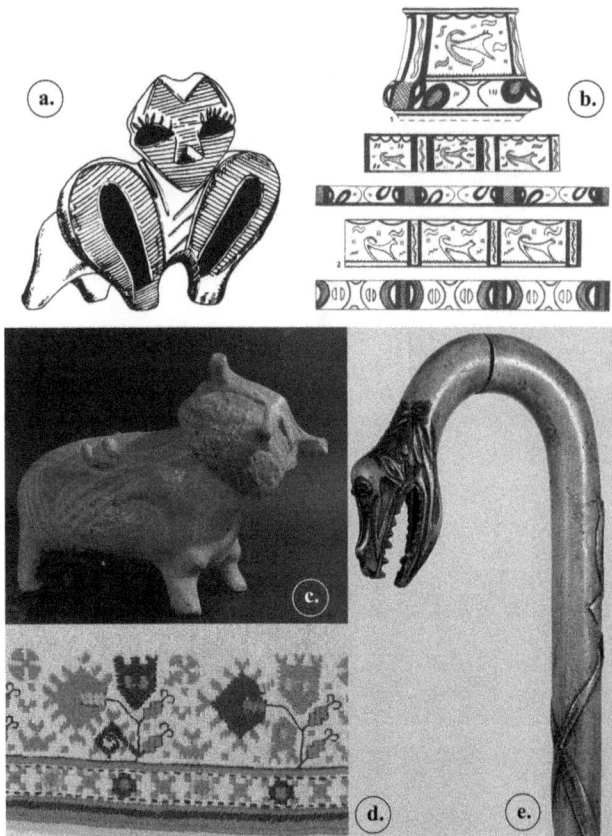

Figure 11. Le dragon combine différents signes d'animaux dangereux, comme les carnassiers, le rapace et le serpent. On le retrouve aussi bien dans des décors peints néolithiques (b) et dans la statuaire (a et c) que dans les crosses de berger actuelles (e). (a) Gimbutas, 1991, p. 180, fig. 279; (b) Gimbutas, 1991, p. 233, fig. 362; (c) Kruta, 1992, p. 94, fig. 70; (d) Kovacheva-Kostadinova et al., 1994, p. 72, fig. 113; (e) Kovacheva-Kostadinova et al., 1994, p. 78, fig. 139

OTHER FACES OF THE MEGALITHIC IN THE NORTH-EAST ALENTEJO – PORTUGAL AND THE REUSE OF TOMBS

Jorge de OLIVEIRA, Clara OLIVEIRA
CHAIA – Univ. of Évora

THE ISSUE

Long gone are the days when the concept of Megalithism only included great megalithic monuments – monuments made using large stones. Later, it was recognised that using this term – megalithism – was particularly limiting because, in truth, examples with large stones are merely a tiny part of an enormous number of displays which also include great dolmens and menhirs, alone or in groups. Although megalithic structures have been the subject of study for a long time, today we question ourselves paradoxically if we should place all these structures in the same cultural bag which was, until now, the megalithic. By recognising, for example, the fact that the large menhirs clearly predate dolmens and the notable recurrent inclusion of the former – which are apparently functionless – in funerary structures compels us to question whether we should include them all in the same ritual package. Strangely, although the most monumental megaliths have caught the attention of all researchers and interested parties since the times of the "Antiquarians" until today, we are still in a sea of doubts.

We must therefore understand that Megalithism – even if we do not understand exactly what it is, since it has myriad forms – seems to appear with the first productive societies and disappears as societies introduce metalwork. It is a complex mythology, whose different ritual forms reflect the socio-economic settings where they take place. Although subject to rules and precepts linked to beliefs which are sufficiently strong and widespread, each community appears to have ritualised the myths in its own way, i.e., in a way linked to its economy and dependent on the geological environment where they take place but always keeping, of course, a coherent link with primordial beliefs.

These specific characteristics are easily recognisable regardless of chronologies, and this is another enormous problem. The chronological variation in the occurrence of megalithic structures makes understanding variations in the structures themselves even more difficult. There is a great tendency, above all among anthropologists, to interpret any cultural display within a progression from simple to complex and it is generally recognised that at the end of a cycle there is a tendency to go back to its origins, with simpler displays similar to those from initial phases. Therefore, if we did not have absolute dates we would be forced to place simpler, smaller megalithic architecture closer to the beginning or end of a cycle, and larger types at its peak. However, at least in the region in question, carbon dating and the type of moveable materials used show that this division is not that clear. To further accentuate the number of doubts, we have to take into account the continued process of reuse which has been observed, above all, in larger scale funerary monuments. We must also recognise that sites related to death have always – as today – been spaces of reference, to be visited and reorganised, and for devotion. Therefore, throughout the projects carried out by us in the north-east Alentejo, specifically in the drainage basin of the River Sever, we have identified a significant number of reuses in monuments which we would now like to present.

Reuse, visits and re-attributing monumental status to funerary spaces complicated even further attempts to identify and understand the rituals which the first agro-pastoral communities developed and which today we generally call Megalithism.

THE SETTING

The North-east Alentejo is roughly defined by the River Sever drainage basin. One of the few rivers which run from south to north, it begins in the north of the Serra of S. Mamede and runs into the River Tagus, its course running for a long way along the Portuguese-Spanish border. The river starts in the quartzite ridges of the mountain range (serra), and its course drains the clay earth of the valley of Aramenha, the long granite strip of Marvão, Castelo de Vide and Valencia de Alcantara and enters schist soils not far from its outlet. Among these different soils, we find different types of megalithic displays, from great menhirs like those in Meada and Carvalhal, large and medium sized granite dolmens in granite areas and small tombs made with schist slabs, naturally located in areas rich in this particular raw material. In the river drainage basin there are currently more than two hundred megalithic tombs, some twenty menhirs, several sites with rock art – both etchings and paintings – which are contemporary with the megalith builders and several inhabited areas. In this region, larger scale funerary architecture are known as passage tombs (antas), or dolmens, while the smaller ones are solely known as graves.

This is the context for the references we bring together here with evidence of the reuse of megalithic monuments.

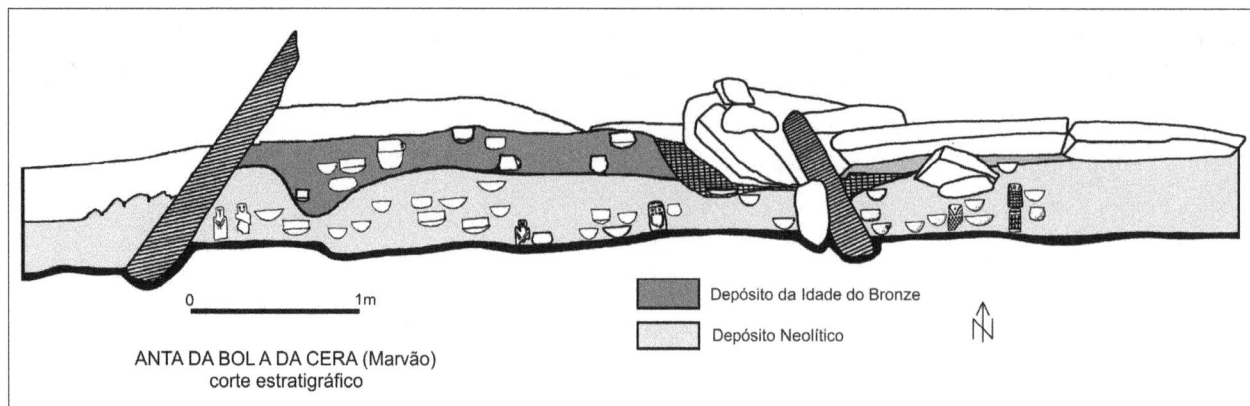

Figure 1. Dolmen of Bola da Cera – Marvão

REUSE

The monuments in the area in question, or at least those excavated by us in the granite areas, showed clear signs that they had been visited over the centuries. If these visits appear to have been motivated at an early stage by the same spirit which led to their construction, in later eras the visits would have been made for other reasons.

The visit made in the Bronze Age to the passage tomb at Bola da Cera (Marvão) (Fig. 1), although with funerary objectives, appears to already show a distinction from the types of visit made before to this and other megalithic tombs. The secondary deposits identified at many monuments, although disturbing earlier deposits in some way, either carried out a general reorganisation of the funerary site, or were merely placed over previous levels. The process used to get to the inside of monuments in later phases is not very clear. Although the method is not known, it seems, however, that secondary tomb builders and contemporary visitors were able to respect the initial architectural structure. The visitors/users to the grave at Bola da Cera – as they were probably driven by other interests, inflicted serious destruction on the funerary structure in order to leave their funerary deposits there.

Removing a support to the side of the chamber and breaking the top of the lead support stone predate the burial which was carried out at this monument. The architectural structure was affected, but the ancient funerary deposits appear not to have been the target of these visitors. A few centimetres below the fractured lead support fragment, two skeletons were found from a first burial with signs of partial cremation. Several objects were found around the skeletons. From the state of preservation of the skeletons and finds, we can be sure that they were not interfered with since they were placed there. It seems, therefore, that the Bronze Age visitors, looking for a place to lay their dead, respected ancient funerary sites, but this concern did not extend to the building itself. The small size of the passage, and especially the contents in the chamber, prevented access to the inside of the monument through its respective passageway. The tumulus would probably be somewhat destroyed and the easiest way to create a tradition

funerary burial site, protected by a large, heavy lid, would be to partially destroy the stone forming the chamber.

Once the burial area was made, the openings were filled in with small stones. At the place where the support was removed, new users built a dry stone wall up to the height of the lid.

Removing, but above all fracturing, one or two chamber supports as a way to gain access to the inside of the monuments is not exclusive to the passage tomb at Bola da Cera. Several examples are known about. The clearest are found in monuments which are still well preserved and especially those which still have a "guillotine stone" at the site. The Marquesa passage tomb (in Valencia de Alcantara) and tomb I at Alcogulo (Castelo de Vide) are monuments which clearly show that breaking side supports was the simplest way to gain access to the inside of the monument.

Megalithic tombs, because of their size and the sites selected, have always attracted different people's attention over the centuries. While in the Bronze Age they were used as a space for tombs, at the time of Roman rule they continued to be visited, excavated and probably even reused as burial sites. The great monument at Tapada de Matos (Castelo de Vide), located next to the Mount of Mosteiros, was clearly part of the main space of a villa known in the area of the mount. During the excavation carried out in the 1980s, it provided a fibula, several coins and pieces made of glass and light sigilata as well as other megalithic finds. At the same site, we discovered that human bones had appeared. With the information we have, we cannot state that any burial was carried out inside the megalithic tomb in Roman times. However, given the presence of coins, a fibula and bones at the same site, we can, at least, consider that hypothesis. Nevertheless, we do know that Romans visited the tomb. We checked that at this monument, during reuse under Roman rule, its new users found some funeral offerings in the dolmen passage when they excavated it. They cautiously dragged the pieces they found to next to the passage's north support. It is a singular set of pieces which could have been coveted by the Romans. They did, however, leave them at the site. This deposit is made up of arrow points, flint blades, schist necklace counters, fragments of pottery and a small zoomorphic sculpture in semi-precious rock.

We also found Roman materials at other monuments. At the Porto Aivado passage tomb, which was totally destroyed and plundered, we found a fragment of tegula in the area which was probably the chamber. Fragments of tegulae were also identified inside the chamber of the tombs at Ribeiro do Lobo (Marvão), S. Gens II (Nisa) and the tumulus of the tomb at Castelhanas (Marvão).

Although some monuments were greatly affected in Roman times and the first centuries of the Middle Ages, others were saved, strangely. The presence of Roman or medieval materials at surface levels of megalithic monuments does not imply that pre-historic funerary spaces were violated. Two interesting examples have been found in the area in question. The Castelhanas tomb, just like the one at Ribeiro do Lobo, both in the county of Marvão, is located in an area which homed an important Late Medieval settlement, under which we can see structures from what was probably Roman villa. At the first excavation carried out, we did not find signs of significant violation and the megalithic structures, although ruined, do not seem to have been used for their stone. Similarly the tombs at S. Gens (Nisa), located in an area with large Roman and medieval settlements, were saved. The presence of some fragments of Roman pottery at surface levels of tomb II at S. Gens does not mean that it was violated.

Under Roman rule and in the first centuries of the Middle Ages we find dolmens, which for reasons unknown were preserved. Deep violations and changes were carried out, however, just before or during the Christian Reconquista, at tombs I and II and probably at tomb III in Coureleiros (Castelo de Vide).

At Anta I in Coureleiros, its violators carried out their actions in the chamber space immediately in front of the corridor. They probably made a fire at this point, and its charcoal has been dated using carbon dating, which showed that it dates to 840+70 years BP (ICEN-592). This violation probably brought about the destruction of the monument's stone structure and the plundering of most of its contents. At tomb III in the same necropolis, a deep violation of the central area of the chamber which penetrated approximately 40cm into the granite soil at the base of the monument. The few finds collected were located around the violation point, next to the supports. At monument II, charcoal recovered next to the chamber entrance, in a stretch of darker ground which disturbed the compacted earth level where some finds were discovered, gave the date 690+130 years BP (ICEN-593).

The visits to dolmens in Coureleiros (Castelo de Vide) dated from the last years of Moorish rule were not the only ones to leave traces behind. At the end of the 15th century or beginning of the 16th, a small house was annexed to tomb IV at Coureleiros (Fig. 2).

Most of the residential structure does not have foundations. The large stones sit on the tumulus. Around the tomb a dry stone wall was built, making it into a pigsty. A somewhat uneven slab over the passage made up a terrace, while the funerary chamber itself was a shelter for animals. To better carried out its function, and since it was already partially destroyed, small dry stone walls linked the supports. On the inside surface of the right hand door frame built over the tumulus, a cross with a triangular base was engraved. Like others found at many different locations, this also appears to Christianise a site which tradition stated had been lived in by Jews. Whether it was inhabited by Jewish people or not, the presence of this cross carved into the door frame of a house attached to and partially built on a a megalithic monument could have a similar meaning to the crosses found on other megalithic monuments, in a clear attempt to make Christian an area which was considered pagan.

The transformation of tomb IV in a pigsty is not unique, either for the time or the area in question. Both on the Spanish and Portuguese sides of the border, several monuments were given this role. Among others, the tombs of Enxeira dos Videias (Marvão), tomb II at Coureleiros (Castelo de Vide), the tomb at Tapada da Anta (Marvão), the tomb at Tapias II and the tomb at Fragoso (Valencia de Alcantara) are examples of the same phenomenon.

Two megalithic graves in the county of Castelo de Vide and one in the county of Marvão were reused for other means. The best example is undoubtedly the tomb at the Mount of Pombal in Castelo de Vide. Plastered and whitewashed with a raised floor, accessed by two steps outside, this interesting dolmen was turned into a pigeon coup at the end of the 19th century. The careful finish, the characteristics of the mortar, the absence of internal niches normally found in pigeon coups, as well as the remains of a skirting board which was first yellow and later blue, and particularly similarities with the tomb chapel of S. Dinis em Pavia, all point towards considering the idea that the monument was turned into a chapel and only later used as a pigeon coup, as we were told at the farmhouse less than 500 metres away.

A chicken run is attached to the south wall of the Monte do Mouratão (Castelo de Vide) site, which today is practically destroyed. The robust building supported by the house was a shelter for poultry. Thick layers of mortar coat the megalithic tomb's chamber almost completely. Only inside can we see the seven supports and the lid which still covers them. An old shepherd who kept sheep and goats in the area was asked about the reasons behind the mount's name, and he told us that he had heard that in the "hut" – as he called the tomb – had lived an old Moorish healer who used the building to practise his science. The mount took its name from the Moor: Monte do Mouratão (Mount of the Moor).

According to Leite de Vasconcelos (Vasconcelos, 1897), the tomb of the Casa do Galhardo, in the county of Castelo de Vide, was related to the figure of the devil.

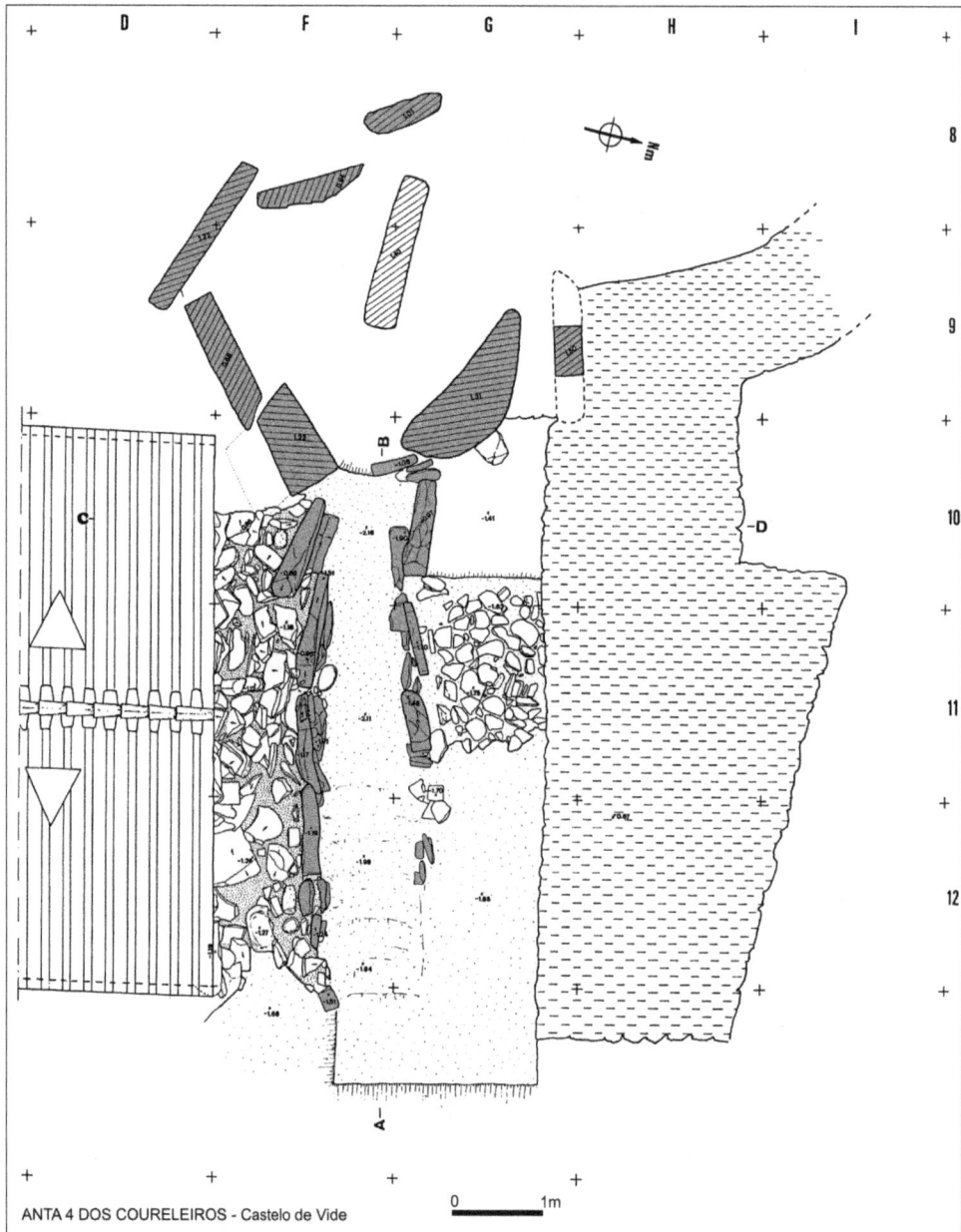

Figure 2. Dolmen 4 of Coureleiros – Castelo de Vide

This author believed that "galh-ardo" meant devil. Nevertheless, no history or legend related to the monument seems to be known in the local area.

Although in the Sever drainage basin we do not know of any tombs which have been clearly turned into chapels, like some found in other areas, there are monuments associated with Catholicism in the area and on the outskirts. In Cedillo, the tomb "Anta del Cabezon" ("Tomb of the large Head") is also known as "Anta Donde Se Reza a La Señora" ("Tomb Where One Prays to Our Lady"). Almost nothing remains of the megalithic tomb. However, there are some ancient ruins nearby which have today been turned into farm houses, and appear to have once belonged to a chapel or church. The name which stuck to the area points towards the existence of a place to pray to the Virgin Mary. Similarly, in the Valencia de Alcantara area, a tomb is located a

few metres from the ruins of the San Anton hermitage. We find a similar story in the county of Nisa. Here, three tombs – one of them in particular – are located a short distance from the chapel of S. Gens. From the examples given, we can see that the process of making sites holy was continued throughout the area over millennia, and places of worship survive to the present day.

In the county of Marvão, tomb II of Meirinha was transformed into a hut (choça). Some of the supports would have been removed to give it a diameter greater than the tomb's chamber. The spaces between the orthostates and their external surfaces were filled in by a thick wall in order to support the wood and the broom which covered it. Although we do not know the orientation of the tomb passage, it is no less interesting to note that the current hut doorway opens to the east.

Choças (huts) are traditional buildings from the region. The circular stone structure is built without mortar and covered by broom and wood, and is cone shaped. These buildings were designed to provide shelter for people and animals.

In the north-east Alentejo, other uses for megalithic tombs are also found. In the area of Cedillo (Spain), a windmill was built on the site of a former passage tomb. Some of its supports were incorporated into the building. At the site where the mill used to stand, today there is only a house in ruins which was probably built using the rubble of the mill. From the tomb we can still see fragments of the supports attached to the east wall of what is left of the house.

The use of dolmens as dividing posts for property in the region is well documented. The tombs at Cabeçuda, Ferrenha, Atalaia, Vale Figueira, Alcogulo I, Coureleiros V, Tapada de Matos, Tapias II, Quatro Lindones, Várzea dos Mourões, Lindon de Campête and others are currently incorporated into walls and were used as reference points in property division.

The buildings of many megalithic tombs may have served many different uses, and their contents was also highly sought after. Searching for legendary treasure that is said to be buried in these monuments would have led to many acts of violation. Tradition states that a box full of golden tools is buried at the Anta do Jardim (Marvão). Two pots could supposedly be found at Salgueirinha (Nisa). One would be full of gold and the other full of poison. If someone searched for the fortune, they could not make a mistake because the wrong pot was opened, they would die instantly. At the tomb in Granja (Marvão), according to tradition, a golden dagger is hidden, left by a horseman from Spain who had been infected by the plague and promised to never use the dagger again if he were cured. When he recovered from his illness, he buried the dagger as promised in the tomb. We should say that although it is a legend, told to us in Beirã, it could contain some truth. Close to the Granja tomb, next to Monte dos Pombais, there is still today a great chimney which was part of a hospital built at the site. Located close to the trail of Saint Mary which led to Spain, this hospital was used to keep people in quarantine, at the time of plagues, when they crossed the border River Sever.

All these legends and traditions led, naturally, to some dolmens been entered and interfered with. And it was not only treasure which people searched for in the tombs. The tradition of placing polished stone (thunderstones) under door thresholds, locks, fireplace or house foundations as a way to protect against lightning was surely responsible for some tombs being gutted in order to obtain such important protection. In the parish of Montalvão, county of Nisa, these polished rock axe heads were also used to forecast the weather. According to older generations, if polished stone axes are thrown into the fire and begin to "sweat and blow", it means that it will soon rain. If they do not change, however, it will remain dry.

As well as collecting axes from dolmens, they were also used as a source of flint for firearms. During the excavations which we led in the Armoury Square of the Castle in Castelo de Vide, we collected several fragments of flint blades adapted for locks on firearms, found at 17th and 18th century levels. The lack of this military device for several periods of war in the region is well documented in a manuscript dating from 12th September 1800, kept at the Manueline Council of Marvão. This document, named "Map showing what is currently needed in Marvão Arms Store, without contemplating what is currently there", refers to "rifle flint", among other military supplies. The lack of flint was satisfied at the time, when, although just before a foreseen war (Guerra de los Naranjos), communications were possible. At other times, however, when the hills were more isolated, the search for "fire stones" must have taken people inside the tombs. In fact, two types of flint fragments were collected from the Arms Store at Castelo de Vide. The larger, thicker fragments, clearly shaped directly for military use, are different from the narrower ones, obtained by breaking up pre-historic blades.

The use of megalithic tombs for the most varied of goals in the north-east Alentejo seems to have taken place continuously, without interruptions. From the reuse of funerary spaces in the Bronze Age to a source of flint for firearms and "lighters", even at the beginning of this century, megalithic monuments have always been visited. Although greatly affected, they have managed to survive until today thanks to the symbolic value which they have held at all times.

References

CANINAS, J. C. Pires and HENRIQUES, F. J. (1985) – *Testemunhos do Neolítico e do Calcolítico no Concelho de Nisa,* in Actas das 1as. Jornadas de Arqueologia do Nordeste Alentejano, Regional Tourism Commission and Castelo de Vide Council, Portalegre.

CANINAS, J. C. Pires and HENRIQUES, F. J. (1987) – *Megalitismo de Vila Velha de Ródão e Nisa,* in Arqueologia no Vale do Tejo, I.P.P.C., Lisbon.

BUENO, Primitiva (1988) – *Los Dolmenes de Valencia de Alcantara,* Excavaciones Arqueologicas en España nº155, Ministry of Culture, Madrid.

BUENO, Primitiva (1989) – *Camaras Simples en Extremadura,* XIX Congreso Nacional de Arqueologia, 1987, Castellón de la Plana.

LEISNER, George and Vera (1956) – *Die Megalithgraber Iberischen Halbinsel. Der Westen (1),* Walther de Gruyter, Berlin.

OLIVEIRA, Jorge de (1997) – *Monumentos Megalíticos da Bacia Hidrográfica do Rio Sever,* 1º Vol., Ed. Colibri, Lisbon.

OLIVEIRA, Jorge (1996) – *"Datas absolutas de monumentos megalíticos da bacia hidrográfica do Rio Sever",* Actas do 2º Congreso de Aqueologia Peninsular, Zamora.

OLIVEIRA, Jorge (1999) – "*Economia e Sociedade dos Construtores de Megálitos da Bacia do Sever*", Actas do 3º Congresso de Arqueologia Peninsular, Vol III, ADECAP, Oporto.

OLIVEIRA, Jorge de (1995) – *Sepulturas Megalíticas del Termino Municipal de Cedillo – Província de Cáceres* – Edicion del Ayuntamiento de Cedillo, Cáceres.

OLIVEIRA, Jorge de (2001) – "*O Megalitismo de Xisto da Bacia do Sever Montalvão – Cedillo*", Muitas antas pouca gente?, Trabalhos de Arqueologia 16, IPA, Lisbon. 2001.

OLIVEIRA, Jorge de (2007) – *The Tombs of the Neolithic Artist-Shepherds of the Tagus Valley*, Actas da I Reunión de Estudios sobre la prehistoria reciente en el Tajo internacional, BAR.

SOME POSSIBLE ASSESSMENTS OF THE DIFFERENT BURIAL SPACES IN THE ALENTEJO AND ARRÁBIDA IN PREHISTORY AND PROTOHISTORY

Leonor ROCHA, Rosário FERNANDES
(CHAIA/ Universidade de Évora)

Abstract: *Prehistorical and Protohistorical funerary customs appear (or appeared) to have a certain regional variation, depending on the geology and geographical area in which they were found. This outlook has been changing in Portugal in recent years as a result of recent excavations carried out in the Baixo Alentejo, showing that exchanges between the coast and inland areas were not only limited to raw materials but also included ideas and models.*

Key Words: *Alentejo; Arrábida; Funerary settings; Prehistory; Protohistory*

Résumé: *Les pratiques funéraires préhistoriques et protohistoriques semble (ou semblait) d'avoir une certaine variation régionale, en fonction de la géologie et lazone géographique dans laquelle ils ont été trouvés. Cette perspectivea changéau Portugal dans ces dernières annéesà la suite defouilles récentes effectuées dans le Bas Alentejo, montrant que les échanges entre la côte etles zones intérieures ne sont pas seulement limitées aux matières premières, mais également des idées et des modèles.*

Mots-clés: *Alentejo; Arrábida; lieux de sépulture; Préhistoire; Protohistoire*

THE AREA

The area included studied in this project is divided into two distinct units, one being Arrábida, the mountain and the sea, and the other the Alentejo, the plains and drylands. They are different in terms of geology and landscape but shared traditions, and as we will see, cultures, in its broadest sense...

Central Alentejo essentially includes the old substrate, a rock formation which in the area is commonly known as *Maciço de Évora* and which is characterised by a great diversity of igneous granite formations and intrusions, as is the case of Reguengos (Carvalhosa, 1983: 206). This geology is located in the Ossa-Morena area, part of the wider peninsular unit known as the Hesperian Massif.

In terms of orography, the Alentejo is dominated by three main phenomena: the Serra d'Ossa to the NW, the Serra do Mendo to the south and the Serra de S. Mamede to the north. They are important control points in the landscape because of the impact they have on the surrounding areas. The presence of prehistorical and protohistorical settlements at the highest points on these hills reflects their strategic importance for these communities.

In terms of topography, the granite areas are usually characterised by moderate relief, thanks to the good conservation of erosion surfaces in the interfluves. They have open, flat-bottomed valleys.

The natural route between the Alentejo and the coast was through a relatively open landscape with gently rolling relief. The area of Évora is, curiously, the contact point between the three main watersheds in the south of Portugal (the Tagus, Sado and Guadiana). It is also the link to two coastal areas with important, unique population centres in Portugal: the Mesolithic shell middens, present in the Tagus and Sado estuaries.

Moving on, we also reach one of the points in Portugal where land reaches furthest into the sea, cape Espichel, where "*the signs of human occupation become rarer and fainter, as if it got less and less as it reached the limits imposed by nature (...). (...) during the day people carried out their ritual obligations but discreetly left it at night to the gods who gathered there: at most they were seen from a neighbouring village at sunset...*" (Ribeiro, 1998: 105).

The mountain range at Arrábida is the most prominent element in the landscape, imposing itself on the land and sea. The mountain range has a length of 35 km, in an ENE-WSW alignment, by 6 km in width, formed by carbonate and marg sedimentary rock, sometimes alternated with detrital units from the Mesozoic era. Other rock formations are found over these ones, above all detrital and carbonate formations from marine environments dating from the Cenozoic era (Ribeiro, 1937; Marçal and Martins, 2005).

Arrábida has a diverse and complex geology with normal faults, tectonic phenomena, deformations, thrust faults and anticlines, among others (Marçal and Martins, 2005).

This mountain chain's heights vary between 500 m, at the top of Formosinho, and 215 m on the Setúbal hills. The Risco range, measuring 380 m, is also important since it is the largest cliff over the sea in mainland Portugal and also the largest carbonate escarpment in Europe (Marçal and Martins, 2005). The surrounding areas have a more gentle relief, from the Pliocene and Quaternary periods (Ribeiro, 1937).

In the carbonate formations in the Western and Middle Borders there is a great number of caves and shelters (Real, 1987).

The two areas have some crops in common even today, including vines and olive trees. In spite of being different in terms of geology and soil, the plant species found there are those best adapted to hot, dry conditions, including the cork oak, holm oak, stone pine, strawberry tree, rosemary, lavander, gum rockrose, agava americana and Indian almond, among others (Medeiros, 1987).

In Arrábida, as in the Alentejo, the trees and dense foliage persisted until the 18th century, where it was full of game (boars, deer, rabbits) and also some wild animals (wolves and bears).

THE TIME: ARRÁBIDA

Funerary settings around the coast are generally very varied structures: natural caves, artificial caves, tombs, *tholoi*.

Some kind of order has been sought for these different building types, either over time, evolution or rock choices. In spite of these archaeological sites showing an undetermined number of uses, where sometimes only the last can be observed or because the discovery and excavation took place prematurely, and that there are often problems in interpreting the layers, some interpretations have been defended, although not without controversy.

For prehistorical funerary structures, a more or less linear evolution sequence has been proposed for the different types of architecture used, in which small monuments without a passage are older, followed by passage tombs, artificial caves and *tholoi* (Cardoso, 1995; Gonçalves, 1999 and 2003; Rocha, 1997 and 2005). Before any megalithic funerary building, however, came the decision to use natural caves, excluding from this set cases such as the cave of Caldeirão (Zilhão, 1992) and funerary settings specific to the early Neolithic era. In fact, the data available for natural caves such as the cave of Cadaval (Oosterbeek, 1995b) or Algar do Bom Santo (Duarte, 1998b) point towards the existence of collective burials which predate the construction of any funerary megalithic monument.

In the transition from the 5th to the 4th millennium BC, and particularly during the first quarter of the 4th, we see the inclusion of structured practices and more symbolic content in funerary settings (Boaventura, 2009), and the number of caves used as necropolises grows from then on. From the 4th millennium BC until the end of the 3rd millennium BC, natural caves were the preferred sites (where they existed) for magic-religious rituals in funerary settings.

In Arrábida, but also throughout the Setúbal peninsula, as well as the presence of the term "Tomb" (*Anta*) in local place names which could indicate the existence of these monuments in the past, there is currently no (confirmed) evidence of the monuments themselves. Neo-Chalcolithic funerary settings appear to be confined only to natural and artificial caves.

In Estremadura, specifically on the Setúbal and Lisbon peninsulas, there are some artificial caves, excavated throughout the 20th century and whose findings point towards their first use coming after natural caves. These are *hypogea*, excavated in soft rocks (generally in so-called soft limestone) which were investigated over the 20th century. They show some architectural polymorphism which varies from a simple cave, accessed through a shaft, up to more complex structures with chambers, anti-chambers and a passage.

On the Setúbal peninsula, three areas with funerary structures with the following characteristics were found: S. Paulo (Almada), made up by two *hypogea*, Convent of Capuchos (Palmela) which had one highly destroyed *hypogeum* and the Quinta do Anjo (Palmela) which had a set of four *hypogea*. (Figure 1).

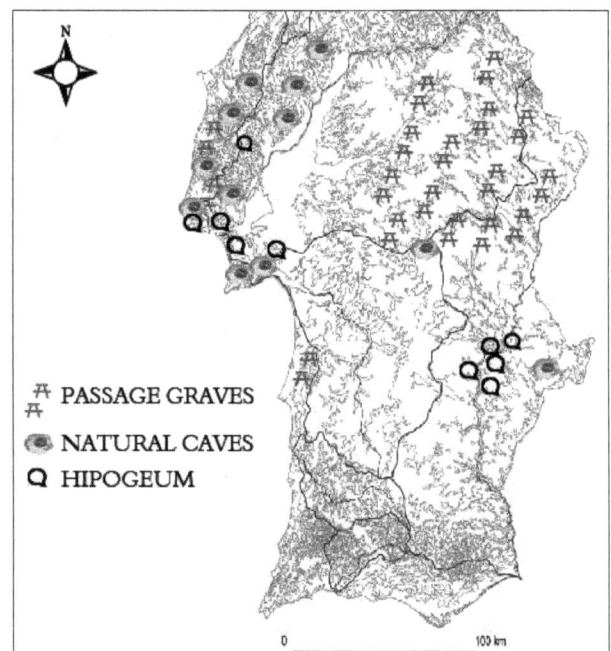

Figure 1. Location of tombs in Southern Portugal

Generally, in Arrábida there are only burials in natural and artificial caves. Naturally this could be the result of population pressure over time, but if we compare it with the Lisbon peninsula, the data which we currently have points towards one situation: megalithic funerary monuments are scarce on the coast.

The differences between coast/inland funerary settings were essentially based on architecture structures, and the presence of some artefacts, such as schist plates,

Figure 2. Natural caves in Arrábida. 1: Cave of Santa Margarida (Setúbal);
2: Cave of Furada (Sesimbra); 3: Cave of Médico (Setúbal)

porcelain plates and crosiers, in Portuguese Estremadura proves, for most researchers, the existence of relationships between the coast and inland areas since at least the 4th millennium. On the other hand, it is still interesting to observe that, in spite of the chronologies of artificial caves not being fully clear, some existing artefacts in this type of tomb point towards a chronology in the second half of the 4th millennium BC which shows the simultaneous use of megalithic monuments and natural caves (Figure 2).

Also, the still scarce analyses carried out on findings in the area, namely the polished-rock axes from Lapa do Bugio (Cardoso, 1992), reveal the origins of raw materials in Alentejo land, although from the coast and lower Alentejo; talc (Odrizola, 2009) may also have originated in the southeast (Huelva).

THE TIME: ALENTEJO

The wealth of Alentejo megaliths has been recognised since at least the 15th century. Over the following centuries, hundreds of monuments were excavated and inventories drawn up by different researchers (Leite de Vasconcellos, Nery Delgado, Carlos Ribeiro, Gabriel Pereira, Emile Cartailhac, Mattos Silva, Filipe Simões, the Leisners, V. Correia, M. Heleno, among others) who would classify their architecture in only four categories: small graves, tombs, *tholoi* and cists.

The first natural cave to be identified as a necropolis (Cave of Escoural) was discovered by chance, only in the second half of the 20th century (Santos, 1964; 1971; Santos *et al.*, 1991; Araújo *et al.*, 1993: Araújo *et al.*, 1995).

As well as excavations, the Alentejo also saw several archaeological prospecting projects which, overall, have significantly altered the range of funerary megalithic monuments known, above all in counties to the West,

with a large increase in the data available. In relation to the coast, the relative abundance of megalithic graves is important, without parallel in other regions, where passage tombs made up almost all monuments (Figure 3).

At the beginning of the 21st century, archaeological work as measures to minimise the impact of large-scale public works (Alqueva dam and related infrastructure) has radically altered the outlook. In fact, a surprising range of new funerary constructions have been identified and excavated. They do not have any parallels in the area but show some similarities with others previously found in the south of Spain and also on the Portuguese coast in the Lisbon and Setúbal areas.

These buildings, a type of *hypogeum* or artificial cave, are excavated in rock and are highly diverse (Figure 4):

– *Hypogea*, with funerary chambers, excavated deeply and are accessed through a shaft, which is almost vertical (Sobreira de Cima – Vidigueira; Outeiro Alto 2 – Serpa);

– Structures excavated in pits or ditches occupying part (Porto Torrão);

– Silo type, surrounded by trenches (Outeiro Alto 2 – Serpa);

– Mine type, with an antechamber and chamber, on slopes (Torre Velha 3; Alto de Brinches 3 – Serpa);

– Simple pits/graves.

This type of individual or mass burial began at the end of the Neolithic period, continuing until at least the first Iron Age. The excellent preservation of some of these burial sites has already allowed researchers to identify woman and children, or couples, embracing (Vinha das Caliças 4, Beja), the latter dating from the first Iron Age.

*Figure 3. Megalithic tombs in Alentejo. 1: Sepultura 7 do Deserto (Montemor-o-Novo);
2: Anta 3ª dos Testos (Arraiolos); 3: Anta Grande do Zambujeiro (Évora)*

*Figure 4. Hipogeum in Southern Portugal. 1: Quinta do Anjo 3 (Palmela); 2: Vale de Barrancas 1 (Beringel);
3: Sobreira de Cima (Vidigueira); 4: Vale de Barrancas 1 (Beringel)*

The findings also have some differences. For older settings, specifically the middle/final Neolithic, there is a strange absence of pottery in sets which include axes, geometrics and blades (Sobreira de Cima and Outeiro Alto 2), unlike what can be found in megalithic monuments dating from the time. At more recent sites, such as Vinha das Caliças, from the first Iron Age, the findings were made up of arms, silver and gold jewellery (necklace beads), bracelets, necklaces, *fibulae*, scarabs, oculated beads, among others, which show links between these populations and the Mediterranean.

Another important new discovery is the fact that many of these funerary buildings, with different chronologies, are found one next to the other, therefore sharing these populations' perfectly defined and well-known grave spaces for millennia. Outeiro Alto 2, for example, has 3 burial centres in pits from the Late Neolithic until the end of the Bronze Age, but with different morphological characteristics.

BETWEEN THE COAST AND INLAND: SOME POSSIBLE ANALYSES

The study of the origins and evolution of funerary sites has passed through several stages, transformed in some cases by a significant increase in available data and in others by stagnation in research into the area itself.

The beginning of the 21st century is a time of great scientific discoveries which question things which were, to some extent, established. In fact, although the issue of megalithic architecture seems to be more or less clear, regional variations notwithstanding, (Rocha, 2005), the same cannot be said of the variations in funerary settings, as seen before.

The more projects are carried out insland and more datings are obtained, the more complex the problem becomes. In fact, this type of structure, which can only be identified in the context of impact minimisation measures since they cannot be identified at the surface, do not stop surprising us with their diversity of shapes and sizes, chronology (from the Neolithic up to, at least, the Roman era) and also by their findings, with great links to the Mediterranean (marble and scarabs).

The differences (and similarities) within the Alentejo and between the Alentejo and the coast, can begin to be classified into different levels:

Architecture

1.1. Similarities between the coast and inland areas in the construction of artificial caves. The layout, with chamber and antechamber, of artificial caves in rock, such as Alapraia, S. Pedro do Estoril (Cascais) and Quinta do Anjo (Palmela), are similar to those at Alto Brinches 3 and Torre Velha 3 (Serpa).

1.2. Similarities between natural caves on the coast and artificial caves inland. The plans of some artificial caves inland, accessed by a shaft, such as Sobreira de Cima (Vidigueira) and Outeiro Alto 2 (Serpa), seem to copy some of the natural caves on the coast, specifically Lapa da Furada (Sesimbra).

1.3. Differences between the north and south Alentejo, with tombs and funerary structures excavated in the rock.

Artefacts

1.4. Similarities between the coast and inland areas, such as the presence of schist plaques, sandstones plaques, crosiers and limestone objects.

1.5. Differences between inland areas and the coast, with marble (animal), amber beads and scarabs in findings.

Burials

1.6. The differences between inland areas and the coast begin to become apparent when considering the conservation of bones. In fact, unlike in tombs, this new type of monument excavated into the rock normally reveal well-preserved bones, which permits anthropological studies and dating to be carried out and to understand that in many of these inland funerary sites, related animals are found. But in particular to analyse the behaviour of these communities towards burials; types of deposition (primary, secondary or transferral), individualisation within a collective space, ossaries and their construction and possible relationship with the management and dynamics of the grave area.

Dating

1.7. Contemporaneity of different type of monument(?). Datings obtained for older burials in *hypogea* in inland areas of the Alentejo (Sobreira de Cima – Vidigueira), have demonstrated high contemporaneity of this type of burial, with those

already known in coastal areas, such as S. Pedro do Estoril 1 (artificial cave), Praia das Maçãs (tholos), in the second half of the 4th millennium BC.

Analysing the map of the south of Portugal allows us to see that the Tagus, Sado and, to a certain extent, the Guadiana, are excellent communication routes between the coast and inland areas, either by boat or over land (Daveau, 1994). These natural paths would have been used, without a doubt, by ancient communities to exchange products but also ideas...Reviewing ancient data from the coast, together with the new discoveries currently being made inland will mean, in the near future, that gaps can be filled, doubts cleared up and a better characterisation made of how these communities moved around the area.

In spite of all the limitations inherent in the fact that many of the inland sites mentioned are still being studied, some interpretations are already possible. On the one hand, it is undeniable that people, ideas and products moved around, showing a clear sharing of social and cultural principles. On the other, the diversity of solutions and influences in areas very close to each other is becoming clearer and clearer, and they cannot be explained only by geological factors, although they can possibly explain the "casing". The variation in findings in adjoining areas and within the same chronology, the longstanding use of grave architecture solutions and/or maintaining the same landscape(defined by the existence of burials from different periods in the same areas) should be clarified with these new data.

The research paths opened with these new factors should show that we are witnessing more continuity and fewer breaks, but that we also see more and more multicultural diversity, which should be fully explained in the future.

References

Monographies

ARAUJO, A.C.; LEJEUNE, M. (1995) – Gruta do Escoural: Necrópole Neolítica e Arte Rupestre Paleolítica. *Trabalhos de Arqueologia*. 8. Lisboa: IPPAR.

AAVV (2009) – *O tempo do Risco. Carta Arqueológica de Sesimbra*. Sesimbra: Câmara Municipal de Sesimbra.

BOAVENTURA, R. (2009) – *As antas e o Megalitismo da região de Lisboa*. Vol I e II. Tese de dissertação de Doutoramento. Lisboa: Faculdade de Letras da Universidade de Lisboa. Policopiado.

CARVALHO, A. M. G. (1968) – Contribuição para o conhecimento geológico da Bacia Terciária do Tejo. *MSGP*. 15. Lisboa: S.G.P.

FERNANDES, R. (2011) – *Da Arrábida ao Alentejo Central, o contributo das grutas no contexto da pré-historia*. Vol. I e II. Tese de dissertação de Mestrado. Évora: Universidade de Évora. Policopiado.

GONÇALVES, V. S. (2005) – *Cascais há 5000 anos. Tempos símbolos e espaços da morte das antigas sociedades camponesas*. Cascais. Câmara Municipal de Cascais.

GONÇALVES, V. S. (2008ª) – *A utilização pré-histórica da gruta de Porto Covo (Cascais): Uma revisão e algumas novidades*. Cascais: Câmara Municipal de Cascais.

RIBEIRO, C. (1880) – *Estudos Prehistoricos em Portugal: Notícia de algumas estações e monumentos prehistoricos. II – Monumentos megalithicos das visinhanças de Bellas*. Lisboa: Typografia da Academia.

RIBEIRO, O. (1998) – *Portugal, o Mediterrâneo e o Atlântico: Esboço de relações geográficas*. Lisboa: Livraria Sá da Costa Editora, 7ª edição revista e ampliada.

SOARES, J. (2003) – *Os hipogeus pré-históricos da Quinta do Anjo (Palmela) e as economias do simbólico*. Setúbal: MAEDS.

SILVA, C.T.; SOARES, J. (1986) – *Arqueologia da Arrábida*. Colecção Parques Naturais. 5. Setúbal.

ZILHÃO, J. (1992) – Gruta do Caldeirão – o Neolítico Antigo. *Trabalhos de Arqueologia*. 6. Lisboa: IPPAR.

Chapters included in monographies

DAVEAU, S. (1994) – A foz do Tejo. Palco de História de Lisboa. *Lisboa Subterrânea*. Lisboa: MNA.

Papers in periodicals

BARROS, L.; SANTOS, P. E. (1997) – Gruta artificial de São Paulo. *Setúbal Arqueológica*. 11-12. Setúbal: MAEDS, p. 217-220.

CARDOSO, J. L. (1992) – A Lapa do Bugio. *Setúbal Arqueológica*. IX-X. Setúbal, Assembleia Municipal de Setúbal, p. 89-225.

CARDOSO, J. L. (1994) – O litoral sesimbrense da Arrábida. Resenha dos conhecimentos da sua evolução quaternária e das ocupações humanas correlativas. *Sesimbra Cultural*. 4. Sesimbra: Câmara Municipal de Sesimbra, p. 5-12.

CARDOSO, J. L.; SOARES, A. M. 1995: "Sobre a cronologia das grutas artificiais da Estremadura portuguesa". *Almadan*. II série. 4. Almada, p. 10-13.

CARDOSO, J. L. (2000) – Na Arrábida, do Neolítico antigo ao Bronze final. *Trabalhos de Arqueologia. Actas do Encontro sobre Arqueologia da Arrábida*. 14. Lisboa: Instituto Português de Arqueologia, p. 45-70.

CARDOSO, J. L.; CARVALHO, A. F. (2008) – A Gruta do Lugar do Canto (Alcanede) e a sua importância no faseamento do Neolítico no território português. *Estudos Arqueológicos de Oeiras – Homenagem a Octávio da Veiga Ferreira*. 16. Oeiras: Câmara Municipal de Oeiras, p. 269-300.

CARVALHO, A. F.; FERREIRA, N. A..; VALENTE M. J. (2003) – A gruta necrópole-neolítica do Algar do Barrão (Monsanto – Alcanede). *Revista Portuguesa de Arqueologia*. 6: 2. Lisboa: IPA, p. 101-119.

CARVALHOSA, A. (1983) – Esquema geológico do Maciço de Évora. *CSGP*. T. 69. Fasc. 2. Lisboa: S.G.P., p. 201-208.

DIAS, M. I.; VALERA, A. C.; LAGO, M.; e PRUDÊNCIO, M.I. (2008) – Proveniência e tecnologia de produção de cerâmicas nos Perdigões. *Actas do III Encontro de Arqueologia do SW (Aljustrel, 2006). Vipasca*. 2, 2ª Série. Aljustrel: Câmara Municipal de Aljustrel.

DUARTE, C. (1998b) – Necrópole neolítica do Algar do Bom Santo. Contexto cronológico e espaço funerário. *Revista Portuguesa de Arqueologia*. 1: 2. Lisboa: IPA, p. 107-118.

DUARTE, C. (2003) – Bioantropologia, Paleoecologia humana e arqueociências. Um programa multidisciplinar para a arqueologia sob tutela da Cultura. *Trabalhos de Arqueologia*. 29. Lisboa: IPA, p. 263-296.

FERNANDES, R.; ROCHA, L. (2008) – Intervenção arqueológica na Lapa dos Pinheirinhos 1 (Sesimbra). *Revista Portuguesa de Arqueologia*. 11. Nº. 2. Lisboa: IPA, p. 29-40.

GONÇALVES, V. S. (2003a) – A anta 2 da Herdade dos Cebolinhos (Reguengos de Monsaraz). As intervenções de 1996/1997 e duas datas de radicarbono para a última utilização da Câmara ortostática. *Revista Portuguesa de Arqueologia*. 6-2. Lisboa: IPA, p. 143-166.

GONÇALVES, V. S. (2008b) – Na primeira metade do 3º milénio a.n.e., dois subsistemas mágico-religiosos no Centro e Sul de Portugal. In: Hernández Pérez, M.; Soler Díaz, J.; López Padilla, J. (coord) – *Actas del 4º Congreso del Neolítico Peninsular*. 2. Alicante: MARQ, p. 112-120.

LAGO, M.; DUARTE, C.; VALERA, A.; ALBERGARIA, J.; ALMEIDA, F. e CARVALHO, A. (1998) – *Povoado dos Perdigões (Reguengos de Monsaraz): dados preliminares dos trabalhos arqueológicos realizados em 1997.Revista Portuguesa de Arqueologia.1*, 1. Lisboa: IPA, p. 45-152.

MONTEIRO, R.; ZBYSZWSKI, G.; FERREIRA, O. V. (1967) – Uma notável placa de xisto encontrada na Lapa do Bugio (Azóia). *Revista de Guimarães*. 77: 3-4. Guimarães, p. 323-328.

MONTEIRO, R.; ZBYSZWSKI, G.; FERREIRA, O. V (1971) – Nota preliminar sobre a lapa pré-histórica do Bugio. *Actas do II Congresso Nacional de Arqueologia (Coimbra, 1970)*. Vol. 1. Coimbra: Junta Nacional de Educação, p. 107-120.

ODRIZOLA, C.; HURTADO PÉREZ, V.; DIAS, M. I. e VALERA, A. C. (2008) – Produção e consumo de campaniformes no vale do Guadiana: uma perspectiva ibérica. *Apontamentos de Arqueologia e Património*. 3. Lisboa: NIA-ERA, p. 45-52.

ODRIZOLA, C. (2009) – Análisis de Procedencia de Campaniformes y una Cuenta de Collar. *O tempo do*

Risco – Carta Arqueológica de Sesimbra. Sesimbra: Câmara Municipal de Sesimbra, p. 156-159.

OOSTERBEEK, L. (1995b) – Elementos para o estudo da estratigrafia da gruta do Cadaval (Tomar). *Al-Madan.* 2ª série. 4-5. Almada, p. 7-12.

OOSTERBEEK, L. (1997a) – Back home! Neolithic life and the rituals of death in the Portuguese Ribatejo. In Bonsall, C.; Tolan-Smith, C. (eds) – The human use of caves. BAR International Series: 667. Oxford: BAR Publishing, p. 70-80.

RIBEIRO, O. (1937) – A Arrábida: Esboço geográfico. *Revista Faculdade de Letras de Lisboa.* 4. I. Lisboa: [s.n.], p. 2-131.

SERRÃO, E. (1958) – Cerâmica Proto-histórica da Lapa do Fumo (Sesimbra) com ornatos coloridos e brunidos. *Zephyrus.* I. Salamanca: [s.n.], p. 176-186.

SERRÃO, E. C. (1975) – Contribuições arqueológicas do sudoeste da Península de Setúbal. *Setúbal Arqueológica.* 1. Setúbal: Junta Distrital de Setúbal, p. 199-225.

SERRÃO, E. C. (1978) – *A Lapa do Fumo. Aspectos e métodos da Pré-História.* Grupo de Estudos Arqueológicos do Porto – GEAP: Porto, p. 27-45.

SERRÃO, E. C. (1978) – Primeiras contribuições para uma periodização do Neolítico e do Calcolítico da Estremadura Portuguesa. *Aspectos e métodos da Pré-História.* Grupo de Estudos Arqueológicos do Porto – GEAP: Porto, p. 17-25.

SERRÃO, E. C.; MONTEIRO, R. (1959) – Estação Isabel: (necrópole pré-histórica da Azóia), Sep.: *I Congresso Nacional de Arqueologia.* I. Lisboa: [s.n.], p. 407-429.

SILVA, C. T.; SOARES, J. (1997) – Chibanes revisitado. Primeiros resultados da campanha de escavações de 1996. *Estudos Orientais.* 6. Lisboa: Instituto Oriental da Universidade Nova de Lisboa, p. 33-66.

VALERA, A.; LAGO, M.; DUARTE, C.; EVANGELISTA, L. (2000) – Ambientes funerários no complexo arqueológico dos Perdigões: uma análise preliminar no contexto das práticas funerárias calcolíticas no Alentejo. *ERA Arqueologia.* 2. Lisboa: ERA/Colibri, p. 84-105.

VALERA, A.; LAGO, M.; DUARTE, C.; DIAS, Mª. I.; PRUDÊNCIO, Mª. I. (2007) – Investigação no complexo arqueológico dos Perdigões: ponto da situação de dados e problemas". *Actas do 4º Congresso de Arqueologia Peninsular.* Faro: Universidade do Algarve.

VALERA, A.; GODINHO, R. (2009) – A gestão da morte nos Perdigões (Reguengos de Monsaraz): novos dados, novos problemas. *Estudos Arqueológicos de Oeiras. 17.* Oeiras: Câmara Municipal, p. 371-387.

VALERA, A. (2010) – Marfim no recinto calcolítico dos Perdigões (1): Lúnulas, fragmentação e ontologia dos artefactos. *Apontamentos de Arqueologia e Património.* 5. Lisboa: NIA-ERA Arqueologia, p. 31-42.

ZILHÃO, J. (1995) – Primeiras datações absolutas para os níveis neolíticos das Grutas do Caldeirão e da Feiteira: Suas implicações para a cronologia da Pré-Historia do sul de Portugal: Kunst, M (coord.) – Origens, estruturas e relações das culturas calcolíticas da Península Ibérica. *Trabalhos de Arqueologia.* 7. Lisboa: IPPAR, p. 113-122.

Electronic doccuments

LAGO, L.; GODINHO, R. (2009) – Notícia do sítio arqueológico do Monte das Covas 3 (Beja). *Apontamentos de Arqueologia e Património – 4 / 2009.* [em linha]. Disponível em: http://www.nia-era.org/index2.php?option=com_docman&task=doc_view&gid=40&Itemid=55.

ROCHA, L. *"As origens do megalitismo funerário no Alentejo Central: a contribuição de Manuel Heleno".* [em linha]. Lisboa: FLL, 2005. Disponível em: WWW: <URL: http://www.crookscape.org/textmar2009/text19.html.

VALERA, A. C.; SOARES, A. M.; COELHO, M. (2008) – Primeiras datas de radiocarbono para a necrópole de hipogeus da Sobreira de Cima (Vidigueira, Beja)", *Apontamentos de Arqueologia e Património – 2/ 2008.* [em linha]. Disponível em: http://www.nia-era.org/index2.php?option=com_docman&task=doc_view&gid=24&Itemid=55.

THE ENEOLITHIC NECROPOLIS FROM SULTANA-MALU ROSU (ROMANIA) – A CASE STUDY

Cătălin LAZĂR

National History Museum of Romania, Bucharest, 030026, Romania
acltara@yahoo.com

Abstract: *The Eneolithic cemetery of Sultana-Malu Roşu is located in the Southeast of Romania and it is one of the most important prehistoric cemeteries currently researched in Romania. From a chronological and cultural point of view it was used by two communities belonging to the Boian and the Gumelniţa cultures and covers the end of the 6th millennium BC and the entire 5th millennium BC. Until now, 50 inhumation graves have been discovered there. The burials are similar in terms of basic elements of the funerary rite. This paper will try to present the funerary traditions of these Eneolithic communities that have used this cemetery, based on the archaeological evidence.*

Key-words: *Cemetery, graves, funerary traditions, Eneolithic, Southeastern Europe*

Résumé: *Le cimetière chalcolithique Sultana-Malu Roşu est situé au sud-est de la Roumanie et il est parmi les plus importantes nécropoles préhistoriques en train d'être recherchées en présent. Le cimetière qui du point de vue chronologique et culturel a été utilisé par les communautés humaines qui appartiennent aux cultures Boian et Gumelnita couvre la fin de VI millénaire BC et tout le V millénaire BC. Les enterrements sont similaires du point de vue des éléments primaires de rituel funéraires. Cette présentation propose de montrer les traditions funéraires spécifiques aux communautés chalcolithiques qui ont utilisé cette nécropole grâce aux donnes archéologiques.*

Mots-clés: *Cimetière, tombes, traditions funéraires, Chalcolithique, sud-est de l'Europe*

INTRODUCTION

The study of funerary practices offers an unique perspective about the past, regarding the funerary rites, habits, spiritual beliefs and social organization of particular human communities, collective and individual identities, traditions and innovations, similarities and differences between cultures, and the presence or absence of interactions between the people of neighbouring and distant cultures. That is why burials have continually exercised a special attraction for anthropologists and archaeologists.

In this context, cemeteries are a special case, because they represent an association of several graves belonging to one or more communities, contemporary or not, constituting an exceptional source of information, throwing light on many aspects of life and death in prehistoric societies (Todorova 1978, 74). Also, cemeteries can provide evidence of specific funerary traditions, the evolution of funerary space, changes and variability of burial behaviour, and finally, but not less important, the identities of past communities.

In this paper we will try to explore the particular case of one of the most recent cemeteries discovered in Romania, Sultana-Malu Roşu, because it has been the subject of complex interdisciplinary research that allowed us to record some original data on specific funerary behaviours of the Eneolithic communities from Southeastern Europe.

GEOGRAPHICAL FRAMEWORK

The Eneolithic cemetery from Sultana-Malu Roşu is located in the northern area of the Balkan region, in the southeast of Romania, on the right bank of the old Mostiştea River (which has been converted into several artificial lakes), about 7 km from the Danube river, near the border with Bulgaria (Fig. 1). From an administrative point of view the site is located in Sultana village, Călăraşi County, Romania (Lazăr et al. 2008, 2009).

The geographical coordinates (Latitude / Longitude) of the cemetery area are 44° 15' 40.3292" N / 26° 52' 2.6103" E, 44° 15' 40.3577" N / 26° 52' 2.8609" E, 44° 15' 39.4114" N / 26° 52' 3.0610" E, 44° 15' 39.1235" N / 26° 52' 1.3720" E, 44° 15' 39.7711" N / 26° 52' 1.2342" E, 44° 15' 39.9275" N / 26° 52' 2.6986" E (Lazăr et al. 2009, 165). The corresponding absolute altitude of this area is at least 45.021 m and maximum 46.740 m. All data are reported in the STEREO-70 projecting system of coordinates and 1975 Black Sea elevation system reference.

From topographical point of view the cemetery is located on the high terrace of the Mostiştea Lake, at 150 m (±1 m) west from the Gumelniţa tell settlement (Sultana-Malu Roşu) and 320 m (±1 m) east from the Boian flat settlement (Sultana-Gheţărie) (Fig. 3).

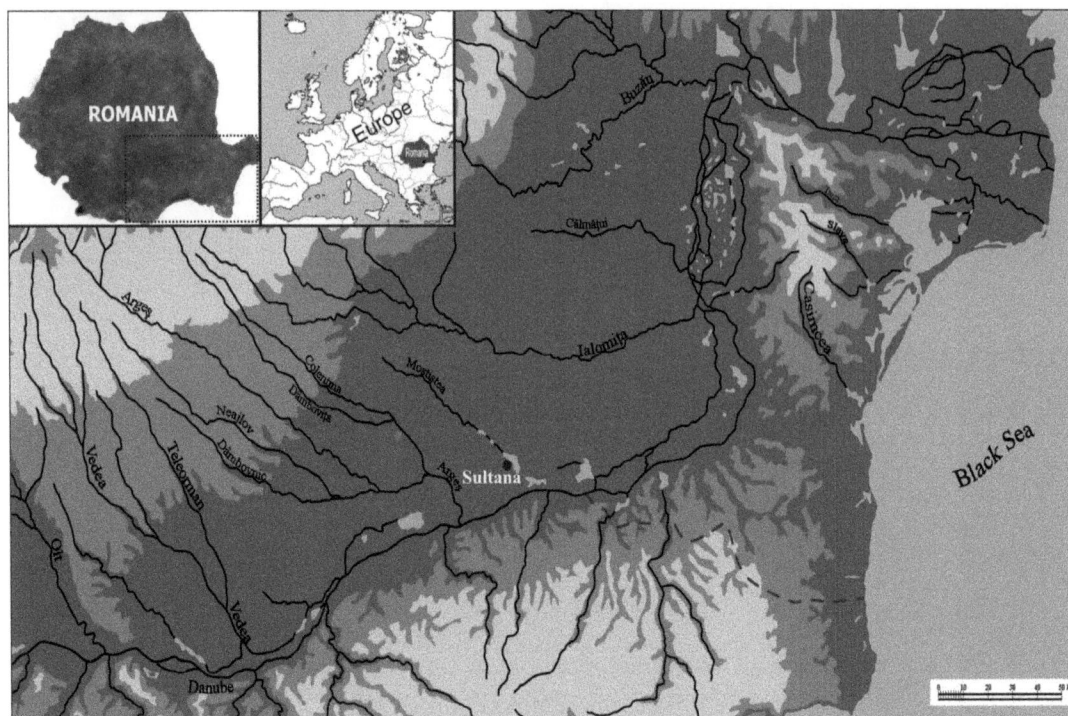

Figure 1. Map of Romania and the location of the Sultana-Malu Roşu cemetery

THE CHRONOLOGICAL AND CULTURAL SETTING

From cultural point of view, the cemetery of Sultana-Malu Roşu was used by two communities belonging to the Boian and the Gumelniţa cultures. Both cultures belong to large Eneolithic cultural complexes, Boian-Maritsa-Karanovo V and Kodjadermen-Gumelniţa-Karanovo VI respectively (Fig. 2), which cover almost all the Balkan area (Todorova 1986; Dumitrescu *et al.* 1983; Bojadjiev *et al.* 1993; Petrescu-Dâmboviţa 2001a; 2001b). Generally, Romanian researchers are using the terms Boian and Gumelniţa cultures and not the full denominations of this cultural units.

Most of Romanian archaeologists consider that the Kodjadermen-Gumelniţa-Karanovo VI cultural aggregate originated through the evolution of the Boian-Maritsa-Karanovo V cultural complex (in the case of the Boian culture only the final phases, as the earliest phases belong to the Middle Neolithic). This phenomenon occurred so rapidly that from its origin it can be referred to as a unique culture with regional attributes (Dumitrescu *et al.* 1983; Haşotti 1997; Petrescu-Dâmboviţa 2001b; Popovici 2010).

The general chronology of these two Eneolithic cultural units (Fig. 2) covers the end of the 6th millennium BC, the entire 5th millennium BC and the beginning of the 4th millennium BC (Dumitrescu *et al.* 1983; Bem 2001; Petrescu-Dâmboviţa 2001a; 2001b).

In terms of absolute chronology, based on AMS radiocarbon dating obtained for Sultana-Malu Roşu

cemetery (n = 5), we can estimate that the graves belong to the probable chronological interval range of 5071 – 4450 cal BC (91.8% – 95.4% probability).

From a cultural point of view, this means that the graves can be included in the Vidra and Spanţov phases of the Boian culture, corresponding to the A1 – A2 phases of the Gumelniţa culture (Fig. 2). However, if we consider the cultural sequence represented in the tell settlement of Sultana-Malu Roşu, then in the future it may also be possible to find graves belonging to the B1 phase of the Gumelniţa culture. On the other hand, these radiocarbon data indicate that Sultana-Malu Roşu cemetery was in use for approximately 600 years. Obviously, this is just a preliminary observation; future excavations and new radiocarbon data will bring supplementary explanations about the chronological range of the use of this cemetery.

METHODOLOGY

The Sultana-Malu Roşu cemetery is one of the few cases in Romania identified following a systematic research approach. With two exceptions (Radovanu and Mǎriuţa sites), in other cases the Eneolithic cemeteries were discovered by chance, due to erosion processes or due to industrial development projects (Comşa 1990, 104; Lazǎr, Parnic 2007, 136-137). On. the other hand, in Romania, identifying the cemeteries was not a priority for archaeologists, the research focusing on settlements, which was in fact a methodological flaw.

In these circumstances, the methodology used at the Sultana-Malu Roşu cemetery was a special one, taking into account the size of the terrace (about 3.5 ha) and the

Figure 2. The Eneolithic chronological and cultural sequence
present at the Sultana-Malu Roşu cemetery

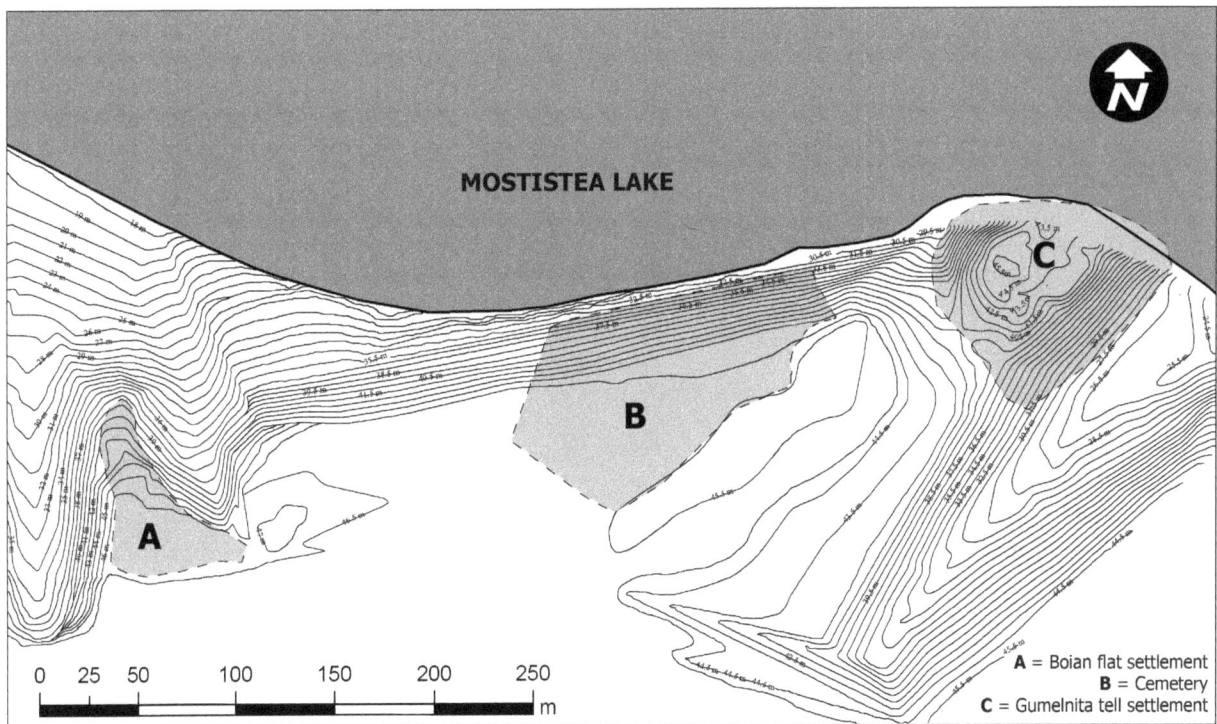

Figure 3. The location of the Sultana-Malu Roşu cemetery and the two settlements

particular aspects posed by the research of a prehistoric funerary areas (Fig. 3). Thus, in 2003, a series of geomagnetic prospects were conducted on the terrace, near the Sultana-Malu Roşu tell-settlement in order to identify the cemetery. After this, in 2007, the site benefited from an investigative aerial research. Corroborated with older aerial photographs, this has led to results regarding the evolution of the landscape, the evolution of the archaeological diggings over time, the erosion of the

lakes banks and the presence of anthropic interventions (Lazăr et al. 2008, 132-133).

The initial excavation method consisted in digging small sections (3 x 1 m or 2 x 1 m), placed at 10-20 m one from another, in order to cover as much terrace surface as possible. After the first graves were uncovered, bigger sections and surfaces were made (8 x 2 m or 10 x 2 m), for systematic and complete research of the targeted

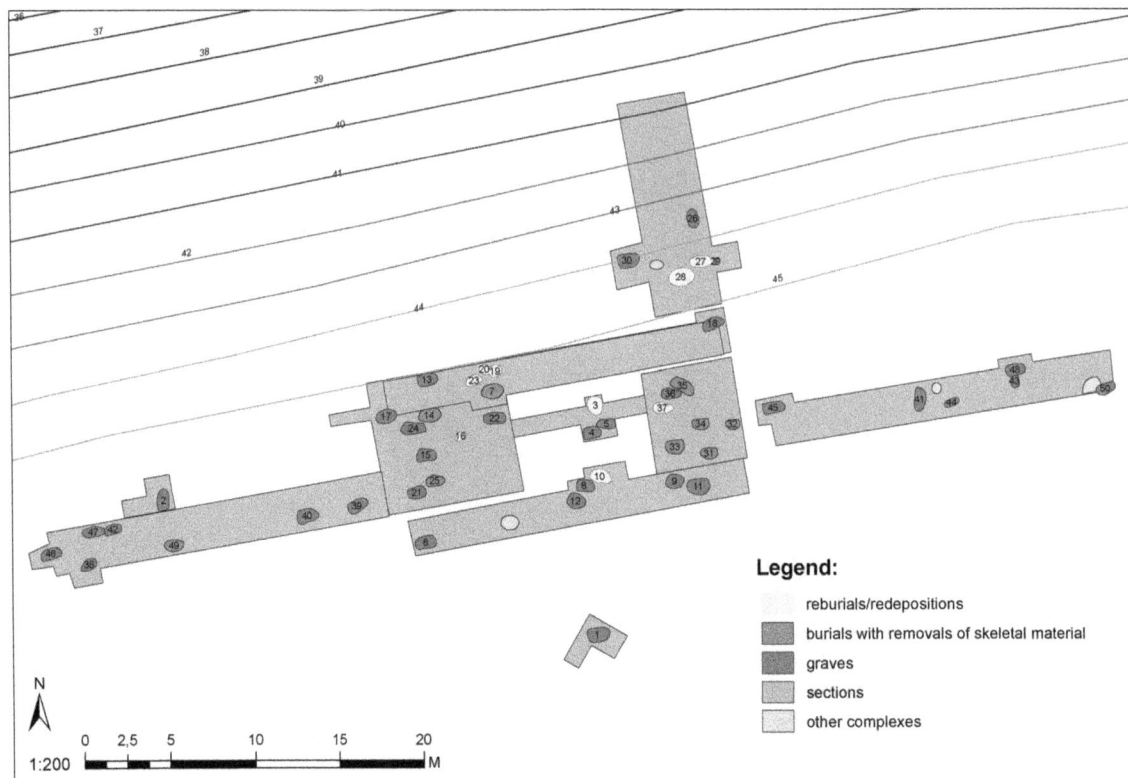

Figure 4. The general plan of the Sultana-Malu Roşu cemetery

areas. Microstratigraphic method was used to record the stratigraphic data, by a thorough analysis of the stratigraphical units (s.u.) (Lazăr *et al.* 2009, 166).

The integration of the archaeological, stratigraphical and topographical data has been achieved through a GIS software.

ENEOLITHIC FUNERARY TRADITIONS IN THE SULTANA-MALU ROŞU CEMETERY

The Sultana-Malu Roşu cemetery is a typical Eneolithic extramural cemetery, with 50 graves discovered until now (Fig. 4). It was used by two settlements belonging to two different communities (Boian and Gumelniţa cultures) (Fig. 3). This funerary area it is placed in proximity of the settlements, in the middle-distance visibility range according to Higuchi's visibility indices, which shows that the necropolis and the settlement are always inter-visible (Lazăr 2011a). Based on the archaeological information and radiocarbon data it is clear that these communities were not contemporary, and therefore there is continuity in the use of cemeteries, which reflects a tradition in the funerary rules of those people and perhaps even descendant communities.

In terms of internal structure and organization, the graves from Sultana-Malu Roşu cemetery are grouped on the terrace edge and slopes, placed at variable distances each other. In some cases they are placed at distances of under 1 m, forming apparent groups (e.g. graves 4 and 5, 8 and 12, 9 and 11, 14 and 24 or 25 and 22), whereas in other

instances graves are more distanced from each other (Fig. 4). Generally, with few exceptions, the graves seem to be aligned into parallel rows oriented to the direction east-north-east. The same organization of the graves was identified in other Eneolithic cemeteries in the Balkans (e.g. Goljamo Delčevo, Targovište, Vărăşti, Devnja etc.) (Todorova-Simeonova 1971; Todorova *et al.* 1975; Angelova 1991; Comşa 1995).

Also, in the perimeter of the cemetery were found four pits (Fig. 4), which have circular shapes, variable dimensions (not very large), and contained pottery fragments, charcoal, burnt clay fragments, stones, animal bones, shells, etc. The presence of such features in the area of the cemeteries may be related to certain stages of the funerary ceremony (possibly remains of the funerary banquet), or may reflect some commemorating ceremonies. Materials from the pits could be the result of dedicated deposition of the deceased. Similar cases have also been identified in the Eneolithic cemeteries from Vinica and Mǎriuţa (Raduncěva 1976; Lazăr, Parnic 2007).

The burials from Sultana-Malu Roşu cemetery are individual, without complex grave structures. They consist of ordinary pits devoid of plaster lining or any traces of related constructions. Generally, funerary pits had irregular oval shape, of varying sizes, depending on the size of the dead bodies. Based on the data collected so far, the graves from here were not marked on the surface, or the markings that were potentially used were made of materials that were not preserved. The orientation of the pits from Sultana-Malu Roşu cemetery was very

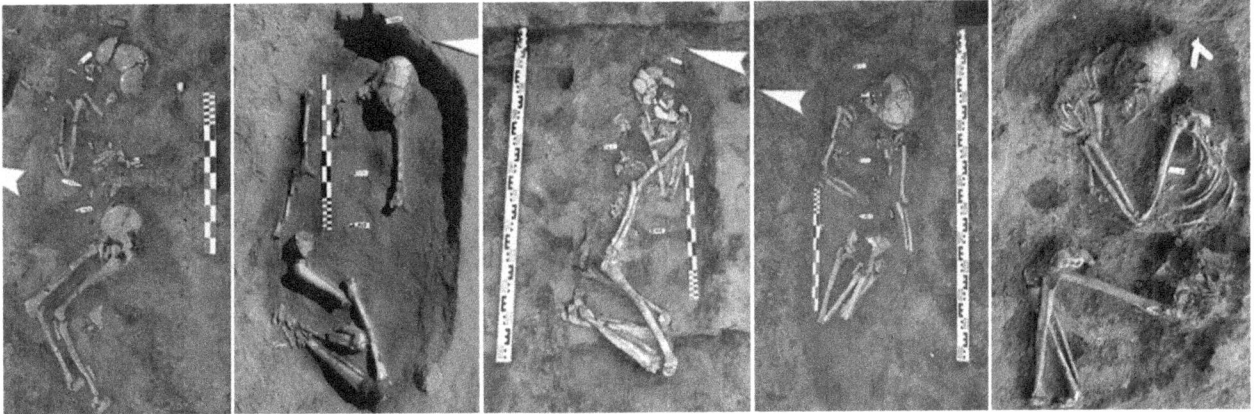

Figure 5. Sultana-Malu Roşu cemetery, examples of primary burials: from left to right, graves no. 24, 17, 18, 22 and 26

Figure 6. Sultana-Malu Roşu cemetery, examples of secondary burials: from left to right graves no. 10, 16 and 28

consistent, along an East – West axis. Only a few cases (graves 2, 3, 26 and 43) had a different orientation, which was closer to a North – South axis (Fig. 4).

The treatment of the dead followed the same pattern, the majority of individuals were laid out in a foetal position (laterally, dorsal or ventral), on the left side, in normal anatomical order (Fig. 5). Only in four cases (graves 22, 24, 26 and 41) the deceased were placed on the right side (Fig. 5). Also, there was one case with the skeleton in ventral extended position (grave 35). The bodies preserved the orientation of the funerary pits (East – West or North – South).

Excepting these cases, at the Sultana-Malu Roşu cemetery are documented also some secondary burials or reburials (Fig. 4, 6). They represent a special situation and until now we have identified nine such cases (graves 3, 10, 16, 19, 20, 23, 27, 28 and 37). Generally, these graves contain human bones from a single individual (Fig. 6). Only in the Grave 28 the long bones from two individuals have been found (Fig. 6). Usually these graves contained the same skeletal elements (long bones, skulls, ribs, vertebrae etc.). Except for the Grave 23, they all contained disarticulated skeletal elements without anatomical connection.

These characteristics, as well as the stratigraphic relationships of these complexes recorded in the field, seem to indicate the intentional deposition of the osteological remains in pits, and on the basis of archaeological data we exclude the possibility of later

intervention. We consider these situations as representing the result of accidental or special circumstances that did not allow for the normal conduction of funerary rites. Alternatively, these complexes may reflect differential treatment of certain individuals of the community, exclusive practices, or occasional habits.

Most of the graves from Sultana-Malu Roşu cemetery are devoid of grave goods. Only 16% of the graves contained material that can be unambiguously assigned to the funeral inventory (Fig. 7). The categories of materials that were found in the graves are represented by pots (grave 6), flint blades (graves 1, 11, 12, 13, 16, 20, 34 and 45), a polished stone axe made of limestone (grave 1) and adornments (graves 1, 13, 14 and 48) represented by beads made of shells of Spondylus gaederopus or Dentalium, bone, marble and malachite (Lazăr et al. 2008, 2009).

So far, we were not able to identify any pattern in the deposition of grave goods or sex and age differences in assigning inventory items.

In Sultana-Malu Roşu cemetery some of the graves (2, 12, 19, 28, 31, 33, and 34) and complexes (C6/2007, C1/2009, C1/2010 and C4/2011) contained animal bones (Fig. 7). These remains were probably placed as offerings. Regarding the animal species, Ovis aries is represented by a metatarsus found in the grave 2; Bos taurus was represented by a diaphysis fragment of a humerus, an almost complete right horncore with a 8 part of the frontal bone attached at the base (C1/2009) and a

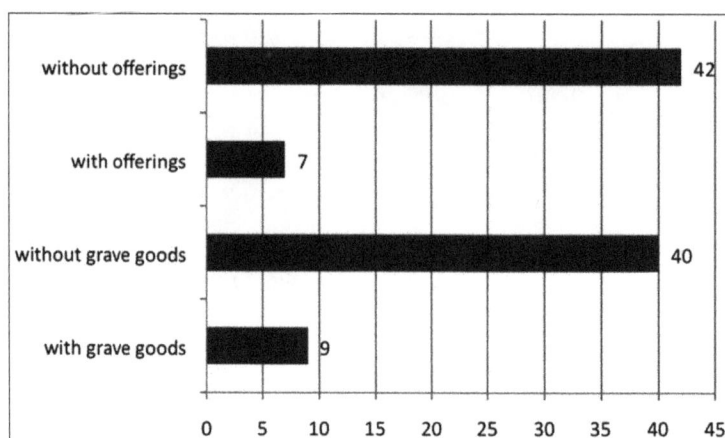

Figure 7. Sultana-Malu Roşu cemetery – grave goods and offerings distribution

neurocranium including a left horncore (grave 28) (Fig. 6). The last two pieces come from the same animal, a fact demonstrating that the two complexes were contemporaneous. Three right upper molars of the same species were found in C6/2007 together with a left lower jaw of Sus domesticus and two Unio sp. shells. Cervus elaphus was represented by a fragment of an un-fused lumbar vertebra found in C1/2009. The same complex contained two other fragments of lumbar vertebra and two diaphyses of long bones which could not be identified taxonomically. Grave 12 contained a river shell and grave 19 an unidentifiable bone fragment. The presence of animal bones within these complexes suggests the practice of a funerary ritual.

Also in the case of funerary offerings we did not identify any pattern of the deposition or sex and age differences in association with these animal bones.

The anthropological analyses made so far on 35 skeletons showed that most of them belonged to adults and only two to Infans (grave 2 and a mandible fragment from grave 21). To them we can also add the three children's graves discovered in 2011 (graves 38, 42 and 43).

There is a higher number of female subjects (n = 22) compared with the number of male subjects (n = 12). We must mention that some of the skeletons (graves 36 – 50) have not yet been anthropologically determined.

In the case of reburials, no preference for male or female subjects was observed, their percentages being equal. From point of view of the age, all individuals belong to the adult category.

Based on the archaeological and anthropological analyses we can conclude that the burial customs do not reflect any sex or age differences.

DISCUSSIONS AND CONCLUSIONS

Generally, most of the Eneolithic cemeteries from the Balkans are associated with a single settlement (Comşa

1974a; 1974b; Todorova 1978; 1982, 1986; Lazăr 2011a). Only in a few cases is a single cemetery associated with multiple settlements, as in the case of our cemetery. In four of those situations we have an association with two settlements (Căscioarele, Durankulak and probably Cernica), and only in two cases probably more than two settlements (Cernavoda and Varna I) (Berciu, Morintz 1957, 83-84, 90-91; Margos 1961, 128-129; 1978, 146-148; Ivanov 1993, 20; Şerbănescu 1996, 28-29; 1998, 14; Comşa, Cantacuzino 2001; Dimov 2002, 28; Slavchev 2010, 206; Lazăr 2011b, 147). All this cases demonstrate that the Sultana-Malu Roşu cemetery is not an exception from the point of view of location and its relationship with the settlements. This situation is very similar to the cemeteries of Durankulak and Căscioarele (Şerbănescu 1996, 1998; Dimov 2002; Lazăr 2011b).

The cemeteries used by different communities, both culturally and chronologically, are best suited for a complex analysis to identify perpetuated traditions and preserved funerary practices. In our case, specific information recorded in this cemetery, indicates the perpetuation of certain funerary rules from the Boian communities to the Gumelniţa communities. This process of perpetuation is not complete and there are also new elements or particular customs. This is a normal situation, especially if we consider that any culture is traditionally seeking to promote their ancestors' rules and knowledge, but also the new ones defined by the coevals (Pouillon 1999).

In terms of grave structure typology the situation from Sultana-Malu Roşu cemetery falls within the series of cases known from other Eneolithic cemeteries, where the funerary pits had irregular oval shape, with varying sizes, depending on the size of the dead bodies (e.g. Durankulak, Vinica, Vărăşti, Goljamo Delčevo, Măriuţa etc.) (Todorova et al. 1975; Radunčeva 1976; Comşa 1995; Boyadžiev 2006; Lazăr and Parnic 2007).

The orientation of the funerary pits along an East – West axis, identified in the most cases from Sultana-Malu Roşu cemetery, is also known in other necropolises belonging to the Eneolithic period from Balkans (e.g. Goljamo

72

Delčevo, Căscioarele, Curăteşti, Radingrad, etc.) (Todorova *et al.* 1975; Ivanov 1982; Şerbănescu 1996; Şerbănescu and Soficaru 2006). This way of positioning the funerary pits may reflect a rule turned into a tradition, transmitted from generation to generation, from the Boian communities to the Gumelniţa communities.

In the case of primary burials, the treatment of the dead followed the same pattern in most of the cases – foetal position (laterally, dorsal or ventral), on the left side (very rarely on the right side), which is a major characteristic of Eneolithic funerary practices from the Balkans. It is very interesting that this type of deposition of the deceased is characteristic both to cemeteries belonging to the Boian culture (e.g. Andolina, Curăteşti, Popeşti-Vasilaţi, etc.), and also for the Gumelniţa communities (e.g. Vărăşti, Targovište, Măriuţa, Vinica, Goljamo Delčevo, Gumelniţa, etc.) (Comşa 1974a, 1974b; 1995; Todorova *et al.* 1975; Raduncčeva 1976; Şerbănescu 1985, 1999; Angelova 1991; Şerbănescu and Soficaru 2006; Lazăr and Parnic 2007; Lazăr 2011b). Most of these cemeteries also contained individuals deposited in a crouched position on the right side, but in a very small percentage.

From the perspective of the graves goods, the typology of the objects identified in the Sultana-Malu Roşu cemetery is similar with the situations encountered in other Eneolithic necropolises from the Balkans. Instead, in terms of wealth, the situation is different, most graves being without inventory. In our opinion, this situation reflects only the research stage and not a past reality.

The secondary burials from Sultana-Malu Roşu cemetery are a rare discovery type for Eneolithic communities from the Balkans. Some similar situations are known in the Eneolithic necropolis of Varna I (burials type 1.D, as catalogued by the author of the excavations), but also in Vărăşti, Vinica and Devnja cemeteries (Todorova-Simeonova 1971; Ivanov 1978; Raduncčeva 1976; Comşa 1995; Chapman 2010). We do not exclude that these cases can be more numerous in the Balkans, but the evidence for secondary burials may be confused by the archaeologist with the disturbance of the graves.

Based on the available information we can conclude that the Sultana-Malu Roşu cemetery corroborates many of the characteristics documented in other cemeteries belonging to the Eneolithic period. From the point of view of the traditions, funerary practices are probably most likely conservative, representing reflections of the eschatological concepts of a particular community accumulated along several generations. On the other hand, they represent extended forms of identities, both personal and collective. Probably during prehistory, as in present, cemeteries were a public space, which made them an ideal place where traditions can be built and rebuilt by the living through the dead. Also, this metaphorical reconstruction allows the living to express individual and group identities. All this demonstrates the existence of some social strategies as social adaptations in the Eneolithic period.

Acknowledgements

We thank Ciprian Astaloş (University College London) for improving the English translation of this paper. This work was supported by a grant of the Romanian National Authority for Scientific Research, CNCSIS – UEFISCDI, project number PN-II-ID-PCE-2011-3-1015.

References

ANGELOVA, I. (1991) – A Chalcolithic Cemetery near the Town of Tărgovište. In J. Lichardus (ed.), Die Kupferzeit als historische Epoche. Symposium Saarbrücken und Otzenhausen 6-13.11.1988, Teil 1. p. 101-105. Bonn: Dr. Rudolf Habelt GmbH.

BEM, C. (2000-2001) – Noi propuneri pentru o schiţă cronologică a eneoliticului românesc. Pontica 33-34, p. 25-121.

BERCIU, D.; MORINTZ, S. (1957) – Şantierul arheologic Cernavoda (reg. Constanţa, r. Medgidia). Materiale si Cercetari Arheologice III, p. 83-92.

BOJADJIEV, J.; DIMOV, T.; TODOROVA, H. (1993) – Les Balkans orientaux. In J. Kozlowski (ed.), Atlas du Néolithique européen. Vol. 1: L'Europe orientale, p. 61-110. Liège. (E.R.A.U.L. 45).

BOYADŽIEV, Y. (2006) – Mobility of individuals and contacts between communities in the 5th millennium B.C. (according to cemeteries information). Bulletin of the Institute of Archaeology XXXIX, p. 13-55.

CHAPMAN, J. (2010) – "Deviant" burials in the Neolithic and Chalcolithic of Central and South Eastern Europe. In K. Rebay-Salisbury, M.L. Sørensen and J. Hughes (eds.), Body Parts and Bodies Whole. Changing Relations and Meanings, p. 30-45. Oxford: Oxbow Books.

COMŞA, E. (1974a) – Istoria comunităţilor culturii Boian. Bucureşti: Editura Academiei RSR. 270 p.

COMŞA, E. (1974b) – Die Bestattungssitten im rumänischen Neolithikum. Jahresschrift für Mitteldeutsche Vorgeschichte für das Landesmuseum für vorgeschichte in Halle-Forschungsstele für die Bezirke Halle und Magdeburg 58, p. 113-156.

COMŞA, E. (1990) – Complexul neolitic de la Radovanu. Călăraşi: Muzeul Dunării de Jos. 121 p. (C.C.D.J. VIII).

COMŞA, E. (1995) – Necropola gumelniţeană de la Vărăşti. Analele Banatului. Serie Nouă IV(1), p. 55-193.

COMŞA, E.; CANTACUZINO, GH. (2001) – Necropola neolitică de la Cernica. Bucureşti: Editura Academiei Române. 254 p.

DIMOV, T. (2002) – Entdeckung und Erforschung der prähistorischen Gräberfelder von Durankulak. In H. Todorova (ed.), Durankulak, Band II. Die Prähistorischen Gräberfelder, Teil 1. p. 24-34. Sofia: Anubis Ltd.

DUMITRESCU, V.; BOLOMEY, A.; MOGOSANU, F. (1983) – Esquisse d'une préhistoire de la Roumanie jusqu'à la fin de l'âge du bronze. Bucureşti: Editura Ştiinţifică şi Enciclopedică. 223 p.

HAŞOTTI, P. (1997) – Epoca Neolitică în Dobrogea. Bibliotheca Tomitana I. Constanţa: Muzeul de Istorie Naţională şi Arheologie. 164 p.

IVANOV, I. (1978) – Les fouilles archéologiques de la nécropole chalcolithique à Varna (1972-1975). Studia Praehistorica 1-2, p. 13-26.

IVANOV, I. (1993) – À la question de la localisation et des études des sites submergés dans les lacs de Varna. Pontica XXVI, p. 19-26.

IVANOV, T. G. (1982) – Tell Radingrad. In H. Todorova (ed.), Kupferzeitliche Siedlungen in Nordostbulgarien, p. 166-174. Materialien zur Allegemeinen und Vergleichenden Archäologie 13. München: C.H. Beck.

LAZĂR, C.; PARNIC, V. (2007) – Date privind unele descoperiri funerare de la Măriuţa-La Movilă. Studii de Preistorie 4, p. 135-157.

LAZĂR, C.; ANDREESCU, R.; IGNAT, T.; FLOREA M.; ASTALOŞ, C. (2008) – The Eneolithic Cemetery from Sultana-Malu Roşu (Călăraşi County, Romania). Studii de Preistorie 5, p. 131-152.

LAZĂR, C.; ANDREESCU, R.; IGNAT, T.; MĂRGĂRIT, M.; FLOREA M.; BĂLĂŞESCU, A. (2009) – New data about the Eneolithic Cemetery from Sultana-Malu Roşu (Călăraşi County, Romania). Studii de Preistorie 6, p. 165-199.

LAZĂR, C. (2011a) – Some Observations about Spatial Relation and Location of the Kodjadermen-Gumelniţa-Karanovo VI Extra Muros Necropolis. In S. Mills and P. Mirea (eds.), The Lower Danube in Prehistory: Landscape Changes and Human Environment Interactions – Proceedings of the International Conference, Alexandria 3-5 November 2010, p. 95-115. Bucureşti: Renaissance.

LAZĂR, C. (2011b) – A Review of Gumelniţa Cemeteries from Romania. Izvestija na Regionalen istoricheski muzej Ruse 14, p. 146-157.

MARGOS, A. (1961) – Otkriti sledi ot novi nakolni celishta vav Varnenskoto ezero. Izvestija na Narodnija Muzej Varna XII, p. 128-131.

MARGOS, A. (1978) – Les sites lacustres dans les lacs de Varna et la nécropole de Varna. Studia Praehistorica 1, p. 146-149.

PETRESCU-DÂMBOVIŢA, M. (2001) – Eneoliticul timpuriu. In M. Petrescu-Dîmboviţa and Al. Vulpe (eds.), Istoria Românilor, vol. I. Moştenirea timpurilor îndepărtate, p. 148-154. Bucureşti: Editura Academiei Române.

PETRESCU-DÂMBOVIŢA, M. (2001b) – Eneoliticul dezvoltat. In M. Petrescu-Dîmboviţa and Al. Vulpe (eds.), Istoria Românilor, vol. I. Moştenirea timpurilor îndepărtate, p. 154-168. Bucureşti: Editura Academiei Române.

POPOVICI, D. (2010) – Copper Age Traditions North of the Danube River. In D. W. Anthony and J. Y. Chi

(eds.), The Lost World of Old Europe. The Danube Valley, 5000-3500 BC, p. 91-111. New York: Princeton University Press.

POUILLON, J. (1999) – Tradiţie. In P. Bonte and M. Izard (eds.), Dicţionar de Etnologie şi Antropologie. p. 673-675. Iaşi: Polirom.

RADUNČEVA, A. (1976) – Vinica – Eneolitno seliste i nekropol. Razkopki i Prouchvania VI. Sofia: Bulgarskata Akademiya na Naukite. 145 p.

SLAVCHEV, V. (2010) – The Varna Eneolithic cemetery in context of the Late Cooper Age in the East Balkans. In D.W. Anthony with J.Y. Chi (eds.), The Lost World of Old Europe. The Danube Valley, 5000-3500 BC. P. 192-210. New York: Princeton University Press.

ŞERBĂNESCU, D. (1985) – Vestigii neolitice descoperite la Ulmeni. Cultură şi Civilizaţie la Dunărea de Jos I, p. 25-35.

ŞERBĂNESCU, D. (1996) – Căscioarele, jud. Călăraşi, Punct C2: D'aia Parte. In Situri arheologice cercetate în perioada 1983-1992. p. 28-29. Brăila: Muzeul Brăilei.

ŞERBĂNESCU, D. (1998) – Căscioarele – D'aia parte, jud. Călăraşi. In Cronica Cercetărilor Arheologice. Campania 1997. A XXXII-a Sesiune Naţională de Rapoarte Arheologice, Călăraşi, 20-24 mai 1998. p. 14. Călăraşi: Muzeul Dunării de Jos.

ŞERBĂNESCU, D. (1999) – Necropola neolitică de la Popeşti, com. Vasilaţi, jud. Călăraşi. In M. Neagu (ed.), Cultura Boian pe teritoriul României, p. 15-16. Călăraşi: Ministerul Culturii.

ŞERBĂNESCU, D.; SOFICARU, A. (2006) – Sultana, com. Mănăstirea, jud. Călăraşi, Punct: Valea Orbului. In Cronica Cercetărilor Arheologice din România. Campania 2005. A XL-a Sesiune Naţională de Rapoarte Arheologice, Constanţa, 31 mai-3 iunie 2006, p. 343-347. Bucureşti: cIMeC.

TODOROVA-SIMEONOVA, H. (1971) – Kusneoeneolitnijat nekropol krai gr. Devnja. Izvestija na Narodnija muzej Varna 7, p. 3-40.

TODOROVA, H.; IVANOV, S.; VASILEV, V.; HOPF, M.; QUITTA, H.; KOHL, G. (1975) – Selishtnata Mogila pri Goljamo Delcevo. Razkopki i Prouchvania V. Sofia: Bulgarskata Akademiya na Naukite. 333 p.

TODOROVA, H. (1978) – The Eneolithic Period in Bulgaria in the Fifth Millenium B.C. British Archaelogical Reports, International Series, no. 49. Oxford: BAR Publishing. 215 p.

TODOROVA, H. (1982) – Kupferzeitliche Siedlungen in Nordostbulgarien. Materialien zur Allegemeinen und Vergleichenden Archäologie 13. München: C.H. Beck. 233 p.

TODOROVA, H. (1986) – Kamenno-mednata Epokha v Bulgariya. Peto Khilyadoletie predi Novata Era. Sofia: Izdatepstvo Nauka i Izkustvo. 280 p.

A GROUP OF OFFERINGS OF COTZUMALGUAPA, GUATEMALA: LATE CLASSIC PERIOD

University of San Carlos of Guatemala, Instituto Politécnico de Tomar /
Universidade de Trás-os-Montes e Alto Douro

Abstract: *The Late Classic (650/700-1000 A.C.) was a remarkable period for the archaeological zone of Cotzumalguapa. During this time the known Cotzumalguapa sculptural style took place. It depicts a variety of iconographic elements showing political content and frequently shows sacrificial themes, among others. In addition, researchers point out some defensive characteristics for that area. This is part of the background in which ritual behaviors such as deposits of offerings are being performed. Offerings found into a small architectural complex located in the archaeological site El Baúl, Guatemala, will be described and to some extent an explanation to the problem is given.*

Key words: *Cotzumalguapa, offerings, Late Classic, archaeology*

Résumé: *La période Classique Tardif (650/700-1000 AC) a été une période remarquable de la zone archéologique de Cotzumalguapa. Pendant ce temps, le style sculptural Cotzumalguapa a eu lieu. Il représente une variété d'éléments iconographiques indiquant le contenu politique et montre souvent des thèmes sacrificiels, entre autres. En outre, les chercheurs soulignent certaines caractéristiques défensives pour ce domaine. Cela fait partie de le contexte dans lequel les comportements rituels tels que le dépôt des offrandes sont en cours d'exécution. Offres trouvées dans un petit complexe architectural situé dans le site archéologique El Baul, le Guatemala, seront décrits et dans une certaine mesure une explication au problème est donnée.*

Mots-clés: *Cotzumalguapa, des offrandes, Classique Tardif, archéologie*

CONTEXT OF THE ZONE OF COTZUMALGUAPA

The zone of Cotzumalguapa is part of the cultural area called Mesoamerica. In this wide area several cultures such as Olmecan, Mayan, Zapotecan, among others, flourished during Prehispanic times. In a general view this period falls between 2000 B.C and 1521 A.D. Specifically the Maya area covers the East of Mexico, as well as Belize, Guatemala, El Salvador and Honduras. The Maya culture was developed in different environments. Thus in the case of Guatemala, traditionally three archaeological divisions are recognized by scholars as following: a) Lowlands toward the North, b) Highlands located approximately in a horizontal strip in the center of the country, and South Coast beside the Pacific Ocean. These areas show differentiated expressions. Thus Cotzumalguapa being part of the South Coast of Guatemala (Figure 1), shows cultural remains noticeably differentiated from those of the well-known lowland Guatemalan archaeological sites such as Tikal, Piedras Negras and Yaxha, among others.

On the other hand, in the specialized literature the term Nuclear Zone of Cotzumalguapa has been defined (Chinchilla 1996) as an area covering approximately 10 square kilometers. It comprises the archaeological sites with monumental architecture, known as El Baúl, El Castillo and Bilbao and its related habitation areas (Figure 2). These sites have special characteristics since they are linked by bridges and causeways (Chinchilla and Carpio 2003: 787).

Archaeological research has showed the area to be inhabited since Pre-classic times, showing a notable cultural development and high population density during the Late Classic (650/700-1000 A.D.). As a result of controlled excavations and casual findings, diverse materials have been recovered and recorded: monumental architecture, collections of pottery, lithic, and others.

A brief exposition of the context of Classic times is given in order to have a general view of the environment where the society inhabited. The archaeological site El Baúl has an area called Acropolis which has an adjacent precinct. These area could be either the residence of an elite group or used for ritual purposes (Chinchilla 1998: 514). The mentioned area is settled on a slope and, in addition, a wall surrounds the precinct. It is believed that these characteristics are related to a defensive nature (Ibid). In the sites of the Nuclear Zone many sculptures have been found, and their style is called Cotzumalguapa. These pieces mostly made of basalt, depict a variety of elements such as animals, humans, mythological beings and others. The sculpture is one of the guidelines to see part of the environment of that time. For instance, some pieces depict abundant scenes or motifs showing skeletal or dismembered beings. Archaeologist Oswaldo Chinchilla who has worked for years in this zone has stated that, sculpture shows many ways of violent conflict which were represented in Mesoamerican art: ballgame, gladiators, human sacrifice (Chinchilla 2009: 140). On the other hand, some offerings have been reported earlier for the zone (Thompson 1948, Parsons 1967). The offerings of Operation EB9 which are the object of this

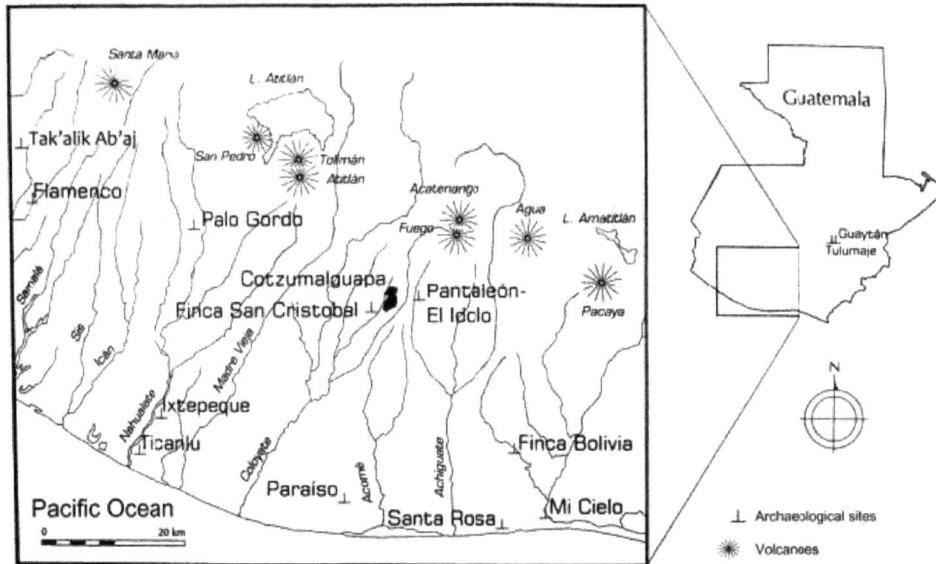

Figure 1. Map showing the Zone of Cotzumalguapa and other archaeological sites of the South Coast of Guatemala

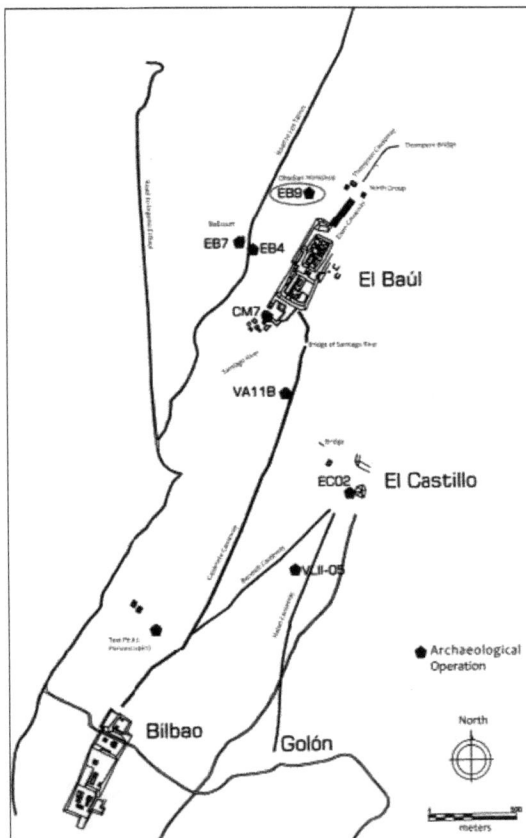

Figure 2. Map of the Nuclear zone of Cotzumalguapa and location of Operation EB9 in the site of El Baúl

study have even been mentioned briefly in several articles as related to an obsidian workshop, and possibly to rituals of purification or birth (Méndez and Chinchilla n.d.; Chinchilla 2006: 5). Most ritual deposits have been associated with the Classic Period and have had often special characteristics, such as combining different pottery shapes and obsidian blades.

SOME PROBLEMS OF INTERPRETATION

Some problems were faced when studying and elaborating an interpretation of the sample. The geographical area of study does not have significant ethno historical data, that can be directly associated with its society. It becomes more difficult when searching for explanations in a religious or ritual field. Other archaeological areas such as the northern Mayan Lowlands or the area of Central Mexico, by contrast, have detailed information of their societies in those areas. This is thanks to reports issued by Spanish friars, for instance, Diego de Landa, Bernardino de Sahagún, Diego Durán, and others. Thus it was necessary to compare archaeological data from similar context, which appear in other zones, together with ethnographic or ethnohistorical descriptions.

THE ARCHITECTURAL COMPOUND AND THE SAMPLE

The items come from an excavated area called Operation EB9 in the archaeological site El Baúl. This operation was located in a space surrounded by a place identified as an obsidian workshop (Chinchilla n.d.), and is approximately 150 meters far from the monumental architectural compound known as the Acropolis. The architecture in this operation shows alignments and other remains of structures of stone which could have had walls of bahareque (Chinchilla 2006: 8-13) (Figure 3). There are two other low structures; one of them is called F. This had remains of a wall which is 0.60 m high. On the other hand, structure G made also of stone appears to have been bigger. Both showed stairs. There were two other structures that are believed to be sweatbaths (Chinchilla 2006: 11-13). These had special traits since they were paved with stones. In addition, a retaining wall was discovered by excavation, on the east side of the

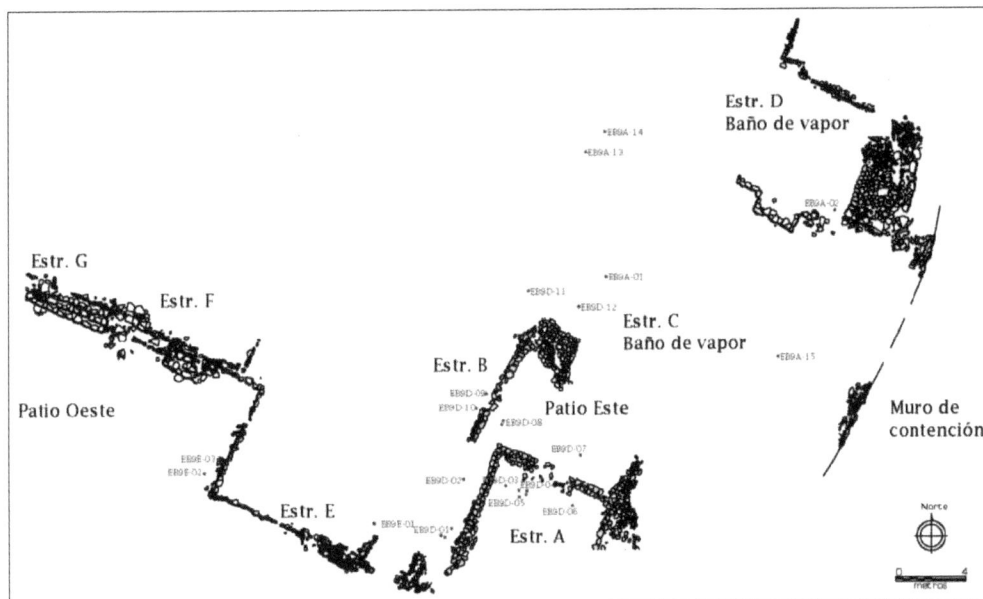

Figure 3. Plan showing the architectural compound of Operation EB9, and location of offerings

architectural group and two spaces were identified as courtyards.

Intentional deposits were found associated with structures and courtyards in this compound (Figure 3). A total 21 offerings consisting of 56 objects were studied (Gómez 2011). Several items were offered below the level of the floor. What remained of the offerings were mainly objects such as, ceramic vessels and obsidian blades. Only one case included a core of obsidian. A small mushroom of basalt was also part of a deposit.

DESCRIPTION AND PROPOSAL OF INTERPRETATION

First, the deposits were classified taking into account material and shape of the ceramic and lithic objects. In order to facilitate comparison a typology was constructed as follows. The number of offerings is indicated in brackets.

Type A: Vase and blade (10) (Figure 4 and 5)

Type B: Plate, blade and core (1)

Type C: Figurine, jar, mushroom stone (1)

Type D: Vase and lid (1)

Type E: Cylindrical Vase (3) (Figure 6)

Type F: Miniature vessels (2) (Figure 7)

Type G: Vessels lip-to-lip (2)

Type H: Pot (1)

When observing the distribution of the offerings in the compound, it is notable how frequent the Type A is in the different spaces. This kind of deposit includes a ceramic vase and an obsidian blade, which is inside the vessel. Types B and C also included an obsidian blade inside a plate and a jar.

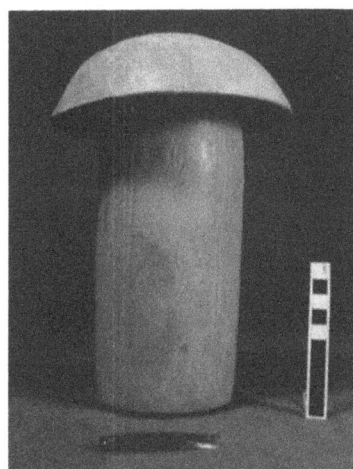

Figure 4. Offering of the Type A. Vase of the ceramic type Plumbate, lid and obsidian blade

Figure 5. Offering of the Type A. Vase of the ceramic type San Andres, showing feet designs, and obsidian blade

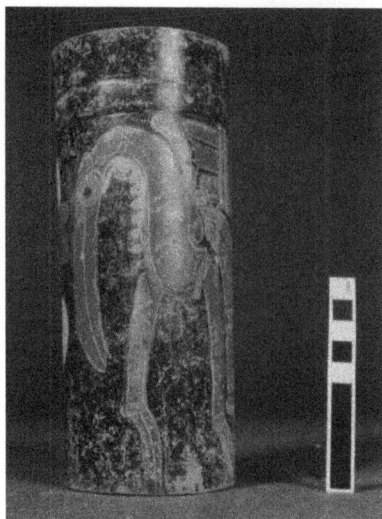

Figure 6. Offering of the Type E, Vase showing a bird

Figure 7. Offering of the Type F. Miniature Vessels

Because of archaeological data from other archaeological sites some comparisons can be done. It was very frequent to find examples similar to those of Type A, B, D, E and G. As the pattern of blade inside a vessel was relatively abundant, it was considered that other contexts that have similar arrangements, could give a guideline to understand some of its function. This pattern is spread over the time and several sites, since Preclassic to Postclassic and from Lowlands to the South Coast. It is important to say that comparisons were done based on bibliographical research which was limited to a sample of sites, mainly of Mayan area, and some of the center and east of México. Some sites that have similarities in that sort of ritual expression are: a) for Lowlands, Tikal, Holmul, Blackman Eddy (Moholy 2008: 9; Coe 1990; Kidder 1947: 21; Mathews and Garber 2004: 52); b) for Highlands, Kaminaljuyú, Salinas de los Nueve Cerros, Baschuc, Nebaj, Santa Cruz Verapaz (Hatch 1997: 16; Dillon *et al.* quoted by Hermes 2003: 31; Becquelin *et al.* 2001: 243-249; Bequelin and Gervais 2001: 81); c) for the South Coast, Tak'alik Ab'aj (Crasborn 2005: 698). Other sites are Yagul, Moxviquil, Teotihuacán (Bernal and Gamio 1974: 36-38; Paris *et al.* n.d.; Elson and Mowbray 2005: 195).

The hypothesis of this work was based on the idea of a public character for the compound, for the abundance of offerings and efforts for its construction, since these do not correspond to a typical domestic area, even when domestic compounds have been less studied in the zone.

In some other archaeological operations in Cotzumalguapa some offerings similar to those of EB9 have been recorded (Chinchilla 1996: 233, 334; Genovéz y Chinchilla 2009; Méndez and Chinchilla n.d.). They are scarce but have been related in some cases, to architecture that is related to public uses. When comparing the offerings of EB9 with cases out of Cotzumalguapa it was visible that deposits of obsidian and ceramic are being earthed below buildings defined as ritual or public, in diverse areas. Some cases however, were possibly domestic (e.g. Yagul) and others were associated to burials (e.g. Salinas de los Nueve Cerros, Baschuc, Nebaj, Teotihuacán).

As in other sites of different periods, the arrangement of one or more obsidian blades inside a vessel has been in some cases, interpreted as dedication to a building, it is possible that in the case of the architectural compound of EB9 the function is similar. This kind of deposits are spread in the architectonic group, and together with the other types could consecrate each structure or space. This idea is supported by looking at ethnographic data that relate a deeply-rooted custom of burying objects, in order to be guaranteed that a new building will work well (Greengberg 1987: 134 quoted by Barabas 2003: 66). An example is related to a domestic context, a group called "chatinos" from the area of Oaxaca, Mexico, has an interesting tradition before constructing a house. In the center of the future house, they excavate a hole where an offering of food is buried. Furthermore in front of the familiar altar the same activity is done. On the 13th day a hole is open near the first one but this time the ritual is dedicated to the entity of fire (Ibid).

In a publication Renfrew and Bahn (1993: 376, 377), give a list of elements that can be useful to identify a specific ritual, and even a deity in an archaeological context. It is important to recognize redundancy of representations in the iconography. In the sample there were some decorated vessels. It was seen a variety of main motifs on them: human; animals such as birds, snake, deer; phytomorphics and other objects, including designs made of lines. This does not make clear the presence of either a mythic concept or other. For this reason, in this paper it is believed that a diversity of intentions and beliefs lead the deposit of offerings. If seen as a whole the space was a place to carry out ritual practices, and probably others, and offerings could work as a way to prepare or consecrate the ground for future rituals. Above the level of offerings some feminine clay figurines were discovered. According to Castillo (2008: 98, 99), they have symbolism and are not simply ornaments. An anthropomorphic figure of clay was also found at the same level, this one has been identified by archaeologist Oswaldo Chinchilla as reminiscence of deities related with skinning (Chinchilla 2006), which are mentioned in ethnohistorical sources of the center of Mexico.

FINAL REMARKS

To conclude, it was evident that more investigation is needed for the area, for different periods to try to understand if there were significant changes in ritual behaviors in this society. As far as it is known and based on analogy with a sample of other sites, it appears that deposits of objects share characteristics since early times and different contexts. What has been discovered until now in the zone of Cotzumalguapa, has characteristics differentiated from other archaeological zones, for example, those offerings including a cylindrical vase which use a bowl or a fragment of vessel as lid. It is also particular that only artifacts of obsidian, mostly blades are in the vessels, when they appear. There are not special materials as happens in other sites, where deposits contain exotic materials (jade, teeth of fauna, shells, among others).

On the other hand, it is possible that blades were used for auto-sacrifice as was recently suggested by archaeologists Luis Méndez y Oswaldo Chinchilla (Méndez and Chinchilla n.d.). Another possibility is that these were symbolic and represented protection for the space where they are. This view is mentioned because of a reference in an ethnohistorical source. Bernardino de Sahagún, friar of the time of the Spanish Conquest, gives information about a superstition in areas of the highlands of Mexico. People believed that a black stone put inside a bowl with water would protect them from warlocks. The bowl was set behind the door or in the courtyard of the house, and when the warlock was seen himself in the water he ran away (Sahagún 1988: Capítulo XXVII, Libro V: 303).

Finally, analysis of contents of vessels and microwear on obsidian artifacts are pending, to have a better understanding of the significance of each offering.

Acknowledgement

Thanks to the European Commission for giving me the opportunity to carry out this research as part of the Master Program Erasmus Mundus in Quaternary and Prehistory. Also thank Instituto Politécnico de Tomar and Universidade de Trás-os-Montes e Alto Douro, Proffessors Davide Delfino, Luiz Osterbeek and Doctor Oswaldo Chinchilla.

References

BARABAS, A. (2003) – Etnoterritorialidad sagrada en Oaxaca. In Diálogos con el Territorio: Simbolizaciones sobre el espacio en las culturas indígenas de México, Vol. 1. p. 37-124. Coordinated by A. Barabas, Serie Ensayos, Col. Etnografía de los Pueblos Indígenas de México CONACULTA-INAH.

BECQUELIN, P. y GERVAIS, V. (2001) – Excavaciones en el valle de Acul y exploración en la cuenca superior del río Xacbal. Arqueología de la región de Nebaj, Guatemala. In Cuadernos de Estudios guatemaltecos 5. p. 25-82, Centro Francés de estudios Mexicanos y Centroamericanos.

BECQUELIN, P.; BRETON, A. y GERVAIS, V. (2001) – Arqueología de la Región de Nebaj, Guatemala.

In Cuadernos de Estudios guatemaltecos 5, Centro Francés de estudios Mexicanos y Centroamericanos.

BERNAL, I.; GAMIO, L. (1974) – Yagul: El Palacio de los seis Patios. Universidad Nacional autónoma de México.

CASTILLO. V. (2008) – Las figurillas moldeadas antropomorfas del Periodo Clásico Tardío de la Costa Sur de Guatemala. Thesis of Licenciatura en Arqueología, Escuela de Historia, Universidad de San Carlos de Guatemala.

CHINCHILLA, O. (1996) – Settlement patterns and Monumental Art at a Major Pre-columbian polity: Cotzumalguapa, Guatemala. Doctoral Dissertation, University of Vanderbilt, Nashville, Tennessee.

CHINCHILLA, O. (1998) – El Baúl: Un sitio defensivo en la zona nuclear de Cotzumalguapa. In XI Simposio de Investigaciones Arqueológicas en Guatemala 1997. p. 512-522, edited by J.P. Laporte, H. Escobedo. Museo Nacional de Arqueología y Etnología de Guatemala.

CHINCHILLA, O. (2006) – Proyecto Arqueológico Cotzumalguapa: Informe de la temporada 2006. Instituto de Antropología e Historia, Guatemala.

CHINCHILLA, O. (2009) – Games, courts, and players at Cotzumalhuapa, Guatemala. In Blood and Beauty: organized violence in the art and archaeology of Mesoamerica and Central America. p. 139-160. Cotsen Institute of Archaeology Press.

CHINCHILLA, O. (s.f.). The Obsidian Workshop of El Baúl, Cotzumalhuapa. p. 203-321.

CHINCHILLA, O. and CARPIO, E. (2003) – El taller de obsidiana de El Baúl, Zona nuclear de Cotzumalguapa. Informe Preliminar. In XVI Simposio de Investigaciones Arqueológicas en Guatemala 2002. p. 787-796, edited by J. P. Laporte, H. Escobedo and H. Mejía. Museo Nacional de Arqueología y Etnología de Guatemala.

COE, W. (1990) – Excavations in the Great Plaza North Terrace and North Acropolis of Tikal. Tikal Report No. 14, Vol. I y II. Serie Editor W. Coe y A. Haviland. Museum Monograph 61. University of Pennsylvania.

CRASBORN, J. (2005) – La obsidiana de Tak'alik Ab'aj en contextos ceremoniales. In XVIII Simposio de Investigaciones Arqueológicas en Guatemala, 2004. p. 695-705, edited by J. P. Laporte, B. Arroyo y H. Mejía. Museo Nacional de Arqueología y Etnología, Guatemala.

ELSON, C. and MOWBRAY, K. (2005) – Burials practices at Teotihuacan in the Early Postclassic period: the Vaillant and Lineé excavations (1931-1932). In Ancient Mesoamerica 16. p. 195-211, Cambridge University Press, USA.

GENOVEZ, V. and CHINCHILLA, O. (2009) – Informe Final: Investigaciones al oeste de la acrópolis de Bilbao, Cotzumalguapa. Proyecto arqueológico Residenciales Santa Lucía, Guatemala (Unpublished).

GÓMEZ, E. (2011) – Las Ofrendas de Cotzumalguapa, Guatemala, en el Clásico Tardío (650/700-1000 d.C.). Master Thesis, Instituto Politécnico de Tomar, Universidade Tras-os-Montes e Alto Douro. Portugal.

HATCH, M. (1997) – Kaminaljuyu/San Jorge: Evidencia arqueológica de la actividad económica en el Valle de Guatemala 300 a.C.-300 d.C. Universidad del Valle de Guatemala.

HERMES, B. (2003) – Vasijas miniatura en las Tierras Bajas Mayas: una visión desde Tikal y Uaxactún. In UTZ'IB, Serie Reportes, Vol. 1 No. 3. Asociación Tikal, Guatemala.

KIDDER, A. (1947) – The artifacts of Uaxactun, Guatemala. Publication 576, Carnegie Institution of Washington, Washington D.C.

MATHEWS, J. and GARBER, J. (2004) – Models of cosmic order. Physical expression of sacred space among the ancient Maya. In Ancient Mesoamerica, 15. p. 49-59, Cambridge University Press, USA.

MÉNDEZ, L. and CHINCHILLA, O. (s.f.). Bajo los pisos: ofrendas y conducta ritual en Cotzumalguapa. Lecture presented at Simposio de Investigaciones Arqueológicas en Guatemala 2010 (In preparation).

MOHOLY-NAGY, H. (2008) – The artifacts of Tikal: Ornamental and Ceremonial Artifacts and Unworked Material. In Tikal Report 27A. University of Pennsylvania Museum of Archaeology and Anthropology, Philadelphia.

PARIS, E.; TALADOIRE, E. and LEE, T. (s.f.). Return to Moxviquil: New investigations and old collections. (In preparation).

PARSONS, L (1967) – Bilbao, Guatemala: An Archaeological Study of the Pacific Coast Cotzumalhuapa Region, vol. 1. Publications in Anthropology, 11. Milaukee: Milwaukee Public Museum.

RENFREW, C. and BAHN, P. (1993) – Arqueología, Teoría, Métodos y Práctica. Translated by María Mosquera, Ediciones Akal, Madrid.

SAHAGÚN, B. Fray (1988) – Historia General de las cosas de Nueva España. Primera versión íntegra del texto castellano del manuscrito conocido como Códice Florentino. Introducción, paleografía, glosario y notas de Alfredo López Austin y Josefina García Quintana. Alianza Editorial Madrid, S. A., Madrid.

THOMPSON, J.E. (1948) – An Archaeological Reconnaissance in the Cotzumalhuapa Region, Escuintla. In Contributions to American Anthropology and History, 44. Washington, D.C., Carnegie Institution of Washington.

ARTISTIC RELATIONS BETWEEN ATTIC VASES PRODUCERS FROM 510 TO 475 B.C. REVIEWED BY THE ATTRIBUTION METHODOLOGY[1]

Carolina Kesser Barcellos DIAS[2]

Museu de Arqueologia e Etnologia da Universidade de São Paulo, Brasil
MAE-USP, Apoio FAPESP

Abstract: *The attribution methodology developed by John D. Beazley is configured as the crucial approach to the study of attic vases. The fundamental principle of this methodology is the recognition of individual hands through a very rigorous stylistic analysis which enables specific traits to be recognized and, finally, gathered in a way so as to assign the producer's work, in other words, the potter's and/or painter's identity. By analyzing the attic black figured vases delineated by context and chronology and attributed to individual artists, our aim is to understand how possible associations, collaborations and interactions between these artists and their workshops occurred from 510 to 475 BC.*

Key-words: *Classical Archaeology, Ceramology, Attribution*

Les rapports artistiques entre les producteurs des vases attiques pendant les années 510 et 475 av. J.-c. revisés a travers la méthodologie d'attribution

Résumé: *La méthodologie d'attribution développée par John D. Beazley figure comme la référence en ce qui concerne les approches d'étude des vases attiques. Le principe de base de cette méthodologie consiste à faire la distinction des mains des artistes individuels grâce à une analyse stylistique rigoureuse qui permette que les traits spécifiques soient reconnus et, enfin, regroupés de manière à permettre la nomination de l'individu producteur, c'est-à-dire, du céramiste et/ou du peintre. Par l'analyse des vases à figures noires délimités contextuellement et chronologiquement, et attribués à des artistes nommés individuellement, nous essayerons de comprendre comment ont eus lieu les possibles associations et coopérations parmi ces artistes dans les ateliers producteurs des vases à figures noires, pendant les années 510 à 475 av. J.-C..*

Mots-clés: *Archéologie Classique, Céramologie, Attribution*

As relações artísticas entre os produtores de vasos áticos durante os anos 510 e 475 a. C. revistas pela metodologia de atribuição

Resumo: *A metodologia de atribuição desenvolvida por John D. Beazley figura como a abordagem determinante para o estudo dos vasos áticos. O princípio básico desta metodologia é a distinção de mãos de artistas individuais através de uma rigorosa análise estilística que permite que traços específicos sejam reconhecidos e, finalmente, agrupados de maneira a possibilitar a nomeação do indivíduo produtor, isto é, do ceramista e/ou do pintor. Através da análise de vasos de figuras negras delimitados contextual e cronologicamente, e atribuídos a artistas nomeados individualmente, procuraremos compreender como ocorreram as possíveis associações e a colaboração entre estes artistas nas oficinas produtoras de vasos de figuras negras, durante os anos 510 a 475 a.C.*

Palavras-Chave: *Arqueologia Clássica, Ceramologia, Atribuição*

Attribution is one of the first and an important step on the recognition of individualities but, more than that, it is a vital tool to more detailed studies about Greek art and craftwork and also about the relations established between their agents in this artistic environment. In this sense, the attribution methodology developed by John D. Beazley[3] became a determinant approach to the study of attic vases: its fundamental principle is the distinction of individual hands of the artists through a rigorous stylistic analysis in which the meticulous observation of the formal details aspects and decoration of the vase allow particular traits to be recognized and, finally, gathered in

---

[1] Article based on the partial results of the post-doctoral research entitled "New perspectives for the attribution: the relationships between artists producers of attic black-figure vases of the Archaic period (510 to 475 B.C.)", developed at the Museum of Archeology and Ethnology of the University of São Paulo, Brazil, between 2010 and 2012, with the support from FAPESP. Constituent communication of the Symposium "Archaeological Dialogues II" – Theoretical and Methodological Reflections 2: Culture Material (SI 07), presented at XVI World Congress of UISPP/XVI Congresso da Sociedade de Arqueologia Brasileira, 04-10 September 2011, Florianópolis, Santa Catarina, Brazil.

[2] I kindly thank Camila Diogo de Souza, Gabriel Castanho de Godoy and Kevin Thierry Cary for the versions and revisions in the English and French language of this article.

[3] Beazley refined the attribution methods through the practice of a systematic observation of styles and graphics of the black and red-figure vases' decoration produced in Athens between centuries VI and IV B.C., and became responsible for the monumental work that constitutes the main source of understanding of diverse aspects of the attic ceramic production. The author published in 1956 the catalog of the painters of 10.000 black-figure vases – Attic Black-figure Vase-painters (ABV). In 1963 he published the three volumes for painters of 21.000 red-figure vases – Attic Red-figure Vase-painters (ARV) and, in 1971, the supplement – Paraliponema (Para) for red and black figures. Beazley's lists present no less than 200 black-figure painters and almost the double of red-figure painters.

order to assign the producer's work, in other words, the potter's and/or painter's identity.

Thanks to the attribution analysis, a very wide panorama of the attic vases has been drawn enabling many questions and different perspectives of study. The identification and grouping of artists provided important tools to the study about painters' and potters' identity, about style, decoration and techniques and also enabled questions about workshops organization.

In essential texts about artists' attribution, such as Haspels ABL (1936) and Beazley ABV (1956), ideas of collaboration, interaction or association between artists appear very briefly, in between lines. Although, there is a certain difficulty to find the development and continuity of these ideas on current researches developed on Greek pottery studies.[4] Some authors are concerned only about part of this question but they do not demonstrate any truly methodological placement to attribution. As there are few bibliographical production in which it is possible to find a discussion about the method used to reach each individual attribution, the debate about methodology vanishes in reviews about artistic recognition and appears as a mere research instrument to organize the documental *corpus*.

Besides, attribution methodology has been focus of severe criticisms, mainly from authors[5] who consider it as just an artistic individualization technique and that would define all its merits. For some of them, individualization is a problem due to the fact that "attributions draw attention to the individual when scholars should be focusing on wider social issues and movements;" and also because "vase-painters are shadowy figures, with no social or historical reality;" (Oakley, 1999: 287) although there are mechanisms to evaluate social and cultural movements by identifying individual artists as these individuals act and participate in their own society.

Founded, therefore, on attribution methodology, we will try to understand how the possible associations and collaboration between artists in the black-figure vases workshops occurred, from 510 to 475 B.C.

The starting point of this analysis comes from some methodological particularities that need to be discussed initially. At first, the terminology used sometimes seems confused or, at least, not very objective putting some questions about work organization in pottery workshops aside. This terminological problem has been already pointed out by Beazley in the preface of ABV (1956: x), when the author affirms that words like "school", "group", and "workshop" have very similar meaning, even not the same, although he used them as different terms. In addition, we can also mention the use of words like "near" and "manner of" for vases attributed to similar style. Beazley points out that some of his attributions were "misquoted" and explains that he is "conscious that the vases placed under the heading 'manner of' an artist are not always in the same category: the list may include (1) vases which are like the painter's work, but can safely be said not to be from his hand, (2) vases which are like the painter's work, but about which I do know enough to say that they are not from his hand, (3) vases which are like the painter's work, but of which, although I know them well, I cannot say whether they are from his hand or not".

In our opinion, with the attribution of vases in the manner of a painter or his workshop, there was an attempt to bring determined pieces closer to some known artists, but a broader perspective was lost when trying to keep these vases away from other artists. On the other hand, when attributing some vases to the manner of a painter or to his workshop, a situation of collaboration between artists was created and we can understand that one object could have been made by not only one hand and, consequently, its production involved more than one artist.

Associations, collaborations and interactions between some artists were made from the beginning of the attribution process even though the consequences and insights of this association do not seem to be developed. The association between the artists of the period selected appears only in a contextual approach, sometimes in an oversimplified manner, as it is demonstrated by Cécile Jubier (1999:181) in her article when the author observes that black figured attic vases in the late archaic period "nous confronte à la masse parfois confuse des artisans du Céramique", even if these artists are individualized in attribution studies. In her article, Jubier debates particularly the case of Diosfos and Safo painters and it seems a bit contradictory that the same author keeps on referring to the artists as a "masse parfois confuse des artisans".

Clearly, at some point, some vases were closely related or even attributed to more than one artist. Instead of discrediting the attribution, this fact reinforced it, establishing social and artistic relations with real implications on the associations between workshops. Through a systematic approach to these artists, we believe it is possible to develop a study that resumes the importance of attribution methodology, amplifying this field to other perspectives less discussed in the texts, for instance, the organization of contemporary workshops and the cooperation and interaction between artists during the manufacturing of specific vases.

Robert Guy, in the article "Artistic personalities" (1999:141-143), points out lines of questionings to be

---

[4] There are a few bibliographical productions focused on one artist, a workshop or a production. "In a period which can be receded to the 30s, until today, there are around 50 titles dedicated to the specific study of an artist and his production, which is relatively little, considering the 70 years that separate us from the first works with this approach" (Dias, 2009:222).

[5] "Groups of scholars, many of them British from the Cambridge University, have questioned the role of attribution in the study of Greek painted pottery. Some have directly attacked the validity and value of the work of Sir J. Beazley, questioning the rationale for attributing vases to painters and the soundness of his methodology. Others, while accepting his results and the potential usefulness of attribution studies, still see problems" (Oakley, 1999:287).

developed based on the lists of artists established by the method, demonstrating that, until the late 90s, even with the validity of Beazley's work, only a little beyond the identification of artists was made. Among the approaches suggested by the author, the most interesting one lies in the development of the analyses with regard to the structure of the workshops, the relation between the techniques, the iconographic choices and the associations between artists.

To the author, the workshops have been observed as being "loosely structured groups of painters, drawn together more by shape than by style of drawing. More readily understood as a series of intersecting circles within which certain artists maintain considerable mobility" (Guy, 1999:142). His suggestion is that style, forms, technique, decoration, iconography and inscriptions must also be observed as evidence of the cooperation between artists-artisans, a study "ultimately more profitable than focusing on the work of individual painters" (Guy, 1999:142). However, the approach would be more thorough without the reservation: the individual's knowledge is fundamental to make it possible to extend the analyses to a whole. In this case, the individualized artist and the comprehension of his production offer information which can answer other questions, such as the relation between the artist and his work, the relationship between artists, the relation between workshops, the relationship between these characters in their specific societies, and the relation of these societies with others.

However, there is a lack of specific studies regarding the artists of the period studied here and, therefore, regarding the relationships between them. The idea of collaboration between artists pass by many of the works dedicated to the attic black-figure pottery, yet there are a few publications which adopt this view in an objective manner. About the workshops' organization, there is the work of T. B. L. Webster (1972), in which the analysis of the relationships between different artists working in diverse workshops is done regarding specific problems: the artists' specialty, that is, the potters' and painters' work, and the clientele's participation as determinant of a production.

Webster (1972:1) seeks to determine the size of a workshop through the quantification of artists who worked in it. For the production of black-figures, the author pointed out an increase in the number of artists at the Ceramicus since the mid-sixth century, but the discussion presented raised some problems that already came from the attribution. Again, the terminological question: Beazley presented the distinction between groups and classes, the first referring to painters, the second to the potters, which already contains some inaccuracies itself. "A group of vases is alike in style of painting but not so alike that it can be regarded as the work of a single artist. A class of vases probably is the work of a single potter, but it is not often possible to say what other classes of vases he made" (Webster, 1972:1). In the accounts, Webster included the vases attributed by

Beazley as "manner of", "near", "related to", etc. because "in each case the vase were painted by some other painter or painters" (1972:2, note 1).

Hence, the terminology adopted at the time of the attribution already provides clues for us to understand, or at least infer, that there is, indeed, a practical collaboration in the vases production. Why then does this idea seem so little disseminated in the studies of attic pottery of this period? Perhaps because, as Webster explains, "all boundaries are fluid. Potter was often painter and painter potter, but we can only be certain of this when we have signatures" (1972:41). From this perspective, the functions in a workshop and the artists' collaboration seem, indeed, difficult to visualize. But the author continues his explanation and demonstrates how much is the work shared in a workshop: "On the whole shops seem to have had a steady membership of potters and painters, but there were roving painters, and special arrangements were made for rush orders; painters could be called in from outside and pots could be bought from neighbouring shops for the home painters to decorate" (1972:41).

Therefore, using the methodology of attribution's principles, defining specific characteristics in a systematic manner, we can delimit individualities, artists' hands and name them, even if in a conventional way. But the attributive work does not end here; it must be carried further, and we believe that the best way for that is observing the traits – formal, stylistic, decorative, iconographic – that are shared by more than one hand and visible on determined objects. This way, connections and collaborations between artists can be visualized, and not only in a superficial manner, but also in a practical and consistent way.

In the memory of André Penin

References

BEAZLEY, J. D. (1956) – Attic black-figure vase-painters. Oxford: Clarendon Press.

BEAZLEY, J. D. (1971) – Paralipomena. Additions to Attic black-figure vase-painters and to Attic red-figure vase-painters. Oxford: Clarendon Press.

DIAS, C. K. B. (2009) – The Gela Painter. Formal and stylistic, decorative and iconographical characteristics. São Paulo: Universidade de São Paulo. Doctorate thesis, 2 vols, English version.

GUY, R. (1999) – Artistic personalities. Céramique et Peinture Grecques – Modes d'emploi. Actes du Colloque International, École du Louvre 26-28 avr., 1995), Paris, p. 141-143.

HASPELS, C. H. E. (1936) – Attic black-figured lekythoi. Paris: E. de Boccard.

JUBIER, C. (1999) – Les peintres de Sappho et de Diosphos, structure d'atelier. Céramique et Peinture Grecques – Modes d'emploi. Actes du Colloque International, École du Louvre 26-28 avr., 1995, Paris, p. 181-186.

OAKLEY, J. (1999) – "Through a glass darkly I": some misconceptions about the study of greek vase-painting. Proceedings of the XVth international congress of classical archaeology, Amsterdam, 12-17 jul. 1998, Classical Archaeology towards the 3rd Millenium: Reflections and Perspectives. Allard Pierson, p. 286-289.

WEBSTER, T. B. L. (1972) – Potter and patron in Classical Athens. London: Methuen.

www.ingramcontent.com/pod-product-compliance
Lightning Source LLC
Chambersburg PA
CBHW061009030426
42334CB00033B/3429